Ecological Literature and the Critique of Anthropocentrism

Bryan L. Moore

Ecological Literature and the Critique of Anthropocentrism

palgrave
macmillan

Bryan L. Moore
Department of English, Philosophy,
and World Languages
Arkansas State University
Jonesboro, AR
USA

ISBN 978-3-319-60737-5 ISBN 978-3-319-60738-2 (eBook)
DOI 10.1007/978-3-319-60738-2

Library of Congress Control Number: 2017944593

Cover illustration: "The Course of Empire: Desolation" Reproduced with Permission of the Collection of the New-York Historical Society

Printed on acid-free paper

This Palgrave Macmillan imprint is published by Springer Nature
The registered company is Springer International Publishing AG
The registered company address is: Gewerbestrasse 11, 6330 Cham, Switzerland

For Carol, Patrick, and Sophie

Preface

This book traces the questioning of anthropocentrism in Western literature with attention to some of the key writers at various points over many centuries, from antiquity to the near-present, and especially those who do so from some sort of ecological perspective. Since the questioning of anthropocentrism lies in works written over a large span of time across the world, my interest in the literary critique of anthropocentrism has compelled me to cast my net widely, though I do not presume to have done much more than scratch the surface in the history of this idea. My critical focus lies on the texts themselves and their contexts, but I admit to a bias in favor of a responsible view of our earthly home. If we are going to overcome our global ecological crises, we must reassess our place on the earth and reduce our impact on the planet. Philosophically and spiritually, but especially in our actions, this demands that we revoke an extreme ("hard") anthropocentrism and act accordingly. The revoking of anthropocentrism is not a new idea or as radical a position as some may imagine. Many of the world's greatest writers have already done this and have prepared the way for us. Why does all this matter? The nonanthropocentric heritage in Western literature is I think, a substantial part of the philosophical and artistic bridge required to help us move more responsibly into the later parts of the twenty-first century and beyond.

Literature written over the past century or so is more likely to disavow anthropocentrism than that written before it, but it is inaccurate to assume that all or even most literature written before such and such year or event (say, the discoveries of Copernicus or Darwin) operates absolutely from the premise that humans are the most important species or the only one that matters. In fact, the assumption that modern works are more questioning about anthropocentrism than ones written before the birth of modern science is not completely safe. As represented in the cliché about onion-peeling, the genesis of the rejection of anthropocentrism would appear to be a post-Darwinian reaction, but it also has some connections to the aesthetic of the Sublime that arose in the eighteenth century, and then, a little before that, it appears to be a result of the Enlightenment, yet certain aspects of it appear before the rise of modern science, and more than a few seeds of the idea stretch back well into early American Indian animism, ancient Greek and Roman philosophy, and Eastern religions. This book explores some of this history. I do not, however, suggest that all of the writers I discuss have identical mindsets. I seek a wide set of references in challenging anthropocentrism, but this is not to claim that, say, Seneca, is a "modern." Seneca's world is very different from that of Donne, and the world of Tennyson is very different from that of Jeffers (to cite more or less random figures). Although space does not allow anything like a full discussion of these contexts, I have tried to keep them in mind.

My working title was *Man Is No Measure*, a revision of a very old idea—i.e., "man is the measure"—but more precisely a line from Robinson Jeffers's poem "The Inhumanist": "Man is no measure of anything" (*Collected* 4:264). (I have, by the way, tried to use the terms "man" and "woman" where the texts use that term but "human" otherwise.) I admit up front that my scope is so large that my execution will fail by overlooking some important works and writers, especially the more contemporary ones. This work is, in reality, "notes toward a history of the critique of anthropocentrism in selected Western writers." Misreadings, misunderstandings, misappropriated contexts, and other misses are inevitable. My interest in this topic has led me to places that are by no means ones of expertise, with Spenser's character Mutabilitie on Arno's Hill and into the Roman ruins with Byron and Shelley, to cite a few examples, and I have found ideas that seem kindred in their interrogation of time and ruminations about the limited roles and abilities of humans on different scales, even as, again, the contexts, worldviews, and epistemologies of the writers differ markedly. The writers I discuss are hardly of one mind about most things. If my readings locate some of the relevance of these works toward our place on earth, physically but also in their ethical and aesthetic implications, then I will have met my goal. While I am interested in more than surfaces of a wide array of texts, I am ultimately more concerned with breadth than depth in my attempt to account for the history of an idea in Western literature.

The implications of my earlier book *Ecology and Literature: Ecocentric Personification from Antiquity to the Twenty-first Century* are essentially affirmative: that trope, I argue there, confirms through scientific, literary, philosophical, and religious works that humans are not completely alone in the universe. It is a common trope and certainly a modernistic one, since it subverts the dominant, traditional mindset that humans may act almost any way they choose, that we (along, perhaps, with our pets) are the only species that matters. Yet it is also a very old one, perhaps a remnant of the interrelationships people felt more distinctly when civilization and progress had not yet removed us so thoroughly from our natural environments. Realizations of the idea in real-world policymaking are difficult to locate. The notion that all living things are kindred and that we are thus bound together is not a feel-good sentiment or wishful thinking but an ethical statement based on ecological and biological science, though, again, many writers suggested this idea long before the rise of modern science. Some may prefer the idea this way: we are all God's creatures and are hence all of worth, though more than a few writers I discuss posit that humans are too hopelessly destructive and self-centered to be able to make any valid claims about the positive worth of our species.

The implications of the present book are also, I believe, affirmative, pointed toward a fuller perspective about the place of *Homo sapiens* in the universe, but they may appear, at least in the short term, gloomy and in some cases negative. The idea that we can undo, perhaps are undoing, or cannot help but eventually undo ourselves and all living things along with us, along with the earth that is our biological basis and home, is not alarmism or an obscure theory but a fact. Life can be wiped out gradually, by environmental degradation (global climate change), economical piracy, human-made or natural plagues, religious fundamentalism, or in the fell swoop of nuclear holocaust, never mind, in descending order of likelihood, a super volcano, an earth-bound comet, or malevolent beings from another planet. Similar to the previous book, this one focuses on our capriciousness, the thorough, largely unquestioned anthropocentrism (attached, it sometimes appears, to a global death wish) that regards the earth solely or primarily as a treasure chest for human consumption.

This is not a book about the apocalypse, all of the possible means of apocalypse, apocalyptical works of literature, or even the narrower category of ecological apocalypse. I am interested in literature and a few other works of art, mostly Western, that show a world without or with fewer humans, works that posit that we are not the end of all existence or the center of the universe. For the sake of perspective and grounding, though, as well as inclusiveness, I do at points address apocalyptic visions, secular and religious, in general. This is also not a book about a literature of human mortality or the brevity of life, which is one of the grand themes of all literature through the ages, from Homer to Herrick, Bunyan to Beckett, even though the brevity of human life is a necessary element in the confronting of anthropocentrism. The realization of life's brevity or the understanding that life will go on without us isn't at all the same as having an ecocentric view, though many of the works center on these facts. Making judgments about texts that decenter humans from those that seek to mortify human ambition or (among older texts) express *contemptus mundi* (religious contempt for the world) has been a challenge, and some of my judgments are (and should be) questionable.

Ecocentrism is a viewpoint that decenters the human subject through an understanding of the interrelations of species in natural environments. To counter anthropocentrism is not the same as expressing misanthropy, but there are certainly texts that do both. A nonanthropocentric or antianthropocentric view is not the same as an antihumanist one or one that is counter to theism; though the impulse is to apply a neat binary, many of the chief humanist writers regularly question the centrality of our species. To cite one example, for Sartre there are two types of humanism. One places humanity as the end and asserts its primacy over all other species, while in the other "man is always outside of himself, and it is in projecting and losing himself beyond himself that man is realized" through "pursuing transcendent goals" (51–52). Many other major writers and intellectual figures as diverse as Sophocles, Shakespeare, Goethe, and Marx have espoused the centrality of humans. The final causes, theodicy, and the centrality of humans in God's order were bulwarks for humanism, though a belief in final causes is also central to writers of the romantic era and beyond. Of course, neither do I suggest that all works that avow anthropocentrism are "flawed" or unworthy of reading, enjoyment, praise, and instruction. It is a given that there are countless works of worthy, even great, anthropocentric literature. Nor do I suggest that all texts that call anthropocentrism into question do so from an ecological basis. Some of the texts that do were written well before ecological science was established in the late nineteenth and twentieth centuries. My definition of "ecological literature" is at times rather broad and loose: some of the works I discuss are only tangentially "ecological."

The book would have been impossible without the love and support of my family, Carol, Patrick, and Sophie, and for these I dedicate this token of appreciation to them. I cannot express my full gratitude to my mother and father, who have supported me in every imaginable way over the years. Thanks also for the support of my brother, Robert Moore, and sister, Pam Lilley. Colleagues in the Arkansas State University Department of English and Philosophy—Jerry Ball, Robert Schichler, Gregory Hansen, Jacob Caton, and Shannon Beasley—read parts or all of my manuscript and gave much needed advice. I'm grateful to Ashton Nichols, who read my manuscript and gave encouragement and helpful feedback. Since I teach a heavy load every semester, it has been difficult to find time for writing and research. Thanks to the school administrators and Faculty Research Awards Committee for granting me a sabbatical for the spring of 2013 to complete major work for this project. (As it turned out, I was for most of that semester stricken with a bulging disc in my back and sciatica that made sitting and typing almost impossible, but I was able to complete some crucial reading, usually while standing, often on my backyard deck, doubtless a strange sight to my neighbors.) I typed most of the manuscript for this book at my computer while one, sometimes two, of our cats (Libby and Artie) lay on the desktop between me and the screen, and they were constant reminders that it is not humans but themselves that are the center of the universe. Thanks to Amber Strother, who, as a graduate assistant in 2010 helped locate some of the texts I discuss; she was a member of a graduate class I taught in 2009 called American Apocalypse, and the class helped me think through some of the ideas in my final chapters. Other members of the class were Angelyn Arnold, Eric Baker, Barry Broussard, Melissa Donner, Adam Fraize, Pratap Kattel, Ali Khalil, Maegon Mayes, Beverly Thompson, and Gabriela Varela-Sanchez. I will never forget the examples, knowledge, and advice I have received from my many great teachers over the years both at Arkansas-Little Rock and TCU. I might not have pursued my profession without the encouragement of Michael Kleine, and I remain indebted to him in particular. I am grateful for permission to reprint here, with substantial revision, articles that originally appeared in the journals *Nature and Culture* and *Ecozon@: European Journal of Literature, Culture and Environment*. Thanks to Counterpoint Press for permission to quote an extended portion of Wendell Berry's poem "The Slip" and to the New-York Historical Society for permission to use the cover image, from Thomas Cole's *Desolation*.

Jonesboro, AR, USA Bryan L. Moore

Contents

CHAPTER 1

Introduction: Anthropocentrism, the Anthropocene, and the Apocalypse

Rhetoric and Anthropocentrism

This is a study of the literary variations of a single theme: our anthropocentric assumptions about the world stand upon weak, largely unexamined bases. While most of the writers central in this book are not scientists, their artistry is based on an understanding of the scientific implications for our place in the world. From this grounding, many of these writers induce attitudinal and active responses from readers, and their work reflects real-world situations, especially concerning how people have asserted power in using and abusing land as well as life, including other humans. The idea that literature is persuasive—rhetorical—undergirds my analysis. Antianthropocentric and outright ecocentric statements operate as what Aristotle, in his treatise *Rhetoric*, calls artistic proofs (*pisteis*) in arguing for a change in direction. Inartistic (or non-artistic) proofs are more or less matters of fact—objective data—used to argue a matter. For example, the fact that the average worldwide temperature has risen over the past several decades provides inartistic proof for the reality of global climate change and potentially the logical basis for a critique of the anthropocentric thinking behind it. Artistic proofs, on the other hand, are created by the rhetorician—the speaker, the writer—to argue something (*On Rhetoric* 1355a). In no case do I contend that the works I discuss are *only* pieces of rhetoric, and few of the writers I discuss in subsequent chapters would call themselves "rhetoricians." Almost none of them employ terms from the field, yet they are rhetoricians. All of them operate in a variety of modes and genres and are open to an array of critical readings.

B.L. Moore, *Ecological Literature and the Critique of Anthropocentrism*, DOI 10.1007/978-3-319-60738-2_1

We are all rhetoricians by nature, even if we never employ or acknowledge that term. The twentieth-century American rhetorician Kenneth Burke wrote that we are beings that by nature use symbolic language.[1] Despite the negative connotations in the word "rhetoric" when used by politicians and in popular media, one cannot *not* use rhetoric. It may be used for good or ill, to promote justice or bottom lines, though it also deals with what is ethical and unethical, just and unjust in given situations. Plato believed that philosophy, through the dialectic, could produce black and white decisions about what is good and bad, but his most important student Aristotle was more concerned with contingencies; we often rely not on truth or Truth but on probable truths. Ethics—their discovery and expression through argument—is central to rhetoric. Aristotle believed that *arête* (moral virtue) is central in the making of a good rhetorician, and the first-century CE Roman rhetorician Quintilian wrote that the good rhetorician is a "good man skilled at speaking" (12.1). Another latter-day American rhetorician, Wayne Booth, writes that "the quality of our lives, especially the ethical and communal quality, depends to an astonishing degree on the quality of our rhetoric" (12).

This book concerns a specific *topos*—Greek for "place," as in "place writers go" to make a case for something, usually in an argument, though the term might be applied more aesthetically. Quintilian writes that the *topoi* are "storehouses of trains of thought" (5.10). The *topos* that concerns me here dramatizes, posits, and imagines a world without people, a world with a diminished human population, or a world with a smaller effect on the natural world than the present reality. The idea is delivered by direct appeals, but also by indirection, by poetic images, monologues, and other techniques. The literary history of the development of the idea is rich and fascinating in its variety, in its wide rhetorical breadth based on many different epistemologies and ideologies. It is a position, a place, but it is also an idea carried through tropes—metaphor, personification, irony, metonymy, and others.

The apocalyptic implications of this *topos* are obvious enough, but in this study, apocalypse is not the focus but rather the imagined, possible means to a diminished human role in the world. A rhetorical exigence—"an imperfection marked by urgency," to quote Lloyd Bitzer (6)—that cannot be modified is not rhetorical. Are anthropogenic catastrophes avoidable? Through education, humility, chance, and change, there may be some hope. The fuller realization that we are a small part of the landscape, a minute feature of our planet's past and (likely) future, might alter our views about ourselves. Yet many scientists state that our species will likely face extinction in the future no matter what actions we take, even if we do begin to take our environmental challenges more seriously and as more important than doing good business. Most of these works are only in a limited way what one may call "predictive"—claiming that this or that event will happen in the near or distant future. Some of the works function as what one may term ecological apocalypse, but they operate less as predictive than as warnings: "If we stay on the current road, something like X will occur." At various points in this book, and especially in the last two chapters, I am concerned with the most speculative of genres regarding not only our future but our past—science fiction. The important science fiction pulp editor Hugo Gernsback promoted the genre as predictive in his work, but the notion that the genre is in essence mostly or even largely predictive is inaccurate. The fact that various science fiction writers—including Jules Verne, H.G. Wells, Arthur C. Clarke, Philip K. Dick, and others—have made predictions about future political and technological features of the world that have been realized in reality are mostly ancillary; this by no means define the genre.

Practicing rhetoric, and doing it well, is, as Wayne Booth suggests, as crucial now to people the world over as it has ever been, though the terms rhetoricians use are fluid and virtually synonymous with those used in other fields. We are awash in data, competing rhetorics, ideologies, advertisements, and plain old noise from an increasing number of directions. Rhetoric is both the means to persuasion as well as the means to understand and critique it, and if people will have a say in how we live, rhetoric is central, though, again, we may not use the terms of the discipline. For some reason, the American public in general is resistant to the facts concerning some of the most pressing issues of our day, including climate change. While a large majority of articles in academic journals—97%—agree on climate change as a phenomenon created largely by the carbon pollution of humans, the American public, thanks to corporate noise machines, ignorant (or sly) politicians, and outright ignorance, believe that scientists are widely divided over the matter. As I write, polls indicate that Americans believe that a mere 42% of scientists agree on the reality of climate change (Zeller). Yet there is reason for hope: it appears that an increasing number of Americans are coming to realize that humans are largely responsible for climate change.[2]

Good thinking about our world will, I think and hope, make us think seriously about the interconnections between ourselves and our natural world. Addressing the holocaust of American Indians, Linda Hogan writes, "It began on this continent, with the genocide of tribal people, and with the ongoing war against the natural world. Here is a lesson: what happens to people and what happens to the land is the same thing" (89). In the same collection, she writes, "To dream of the universe is to know that we are small and brief as insects, born in a flash of rain and gone a moment later. We are delicate and our world is fragile. It was the transgression of Galileo to tell us that we were not the center of the universe," an idea that "even in our own time . . . is treacherous" (126). Joseph Bruchac echoes Hogan: "If we see 'the Earth' as the web of life that sustains us, then there is no question that the web is weakened, that the Earth is sick. But if we look at it from another side, from the view of the living Earth itself, then the sickness is not that of the planet, the sickness is embodied in human beings, and, if carried to its illogical conclusion, the sickness will not kill the Earth, it will kill us" ("The Circle").

Anthropocentrism and Ecocentrism

In 1947, Columbia University History Professor Robert Livingston Schuyler presented a speech entitled "Man's Greatest Illusion." This illusion is anthropocentrism:

> There is in all of us a kind of subconscious vanity, though a better name could perhaps be found for it. We cannot bear to think that we are cosmically unimportant. We can, of course, no longer believe, as our forebears believed up to some ten generations ago, that we are, literally, at the center of the physical universe. Nevertheless it is still possible for us to believe that we are at the apex of creation. In what we know about the universe, to be sure, there is no good reason whatever for believing this . . . (47)

Schuyler does not argue from an ecological perspective, but he does tie the idea to education, stating that the "virtue of humility ought to be a major objective of education" (48). It is difficult, he says, to unlearn the notion that everything that has ever happened has done so for us, living at the present moment in time. "We are not," he states, "final in time any more than we are central in space" (50). If people in general were aware of this seemingly obvious fact, would we treat one another—and the world in which we live—differently? I believe that if we had a more complete understanding of our relative smallness, we would not trash our planet, we would be more concerned about the effects of burning fossil fuels, we would do as much as we could to save the few remaining wilderness areas for posterity, and a humbler view of ourselves would lead us to treating one another more fairly.

Since the time of Schuyler's speech, we have only intensified our ravaging of the planet. Over the past forty years, the global wildlife population has been cut in half ("Living Planet"), the result not of natural selection but actions based on the assumption that we matter more than wild animals, land, and bodies of water. About 40% of the earth's surface is devoted to raising food for one species: humans ("Introduction").[3] According to E. O. Wilson, twenty-five places covering 1.4% of the world's land surface contain the last remaining habitats of 44% of the planet's plant species and 35% of its terrestrial vertebrate species (Hallam 246). We find ourselves in a crisis of our own making. Are we aware enough of it? No. Will we find our way out of it?

Later chapters highlight what is, by most standards, a literature and art that is subversive, directly and indirectly, in conceiving of a natural world that exists for itself, not as a thing to principally be owned, exploited, or profited from. These literary works challenge anthropocentrism, the idea, utterly dominant though still largely unexamined, that (according to *Merriam-Webster's Collegiate Dictionary*) consists of "interpreting or regarding the world in terms of human values and experiences." A.C. Grayling writes that to have an anthropocentric view is

> To see everything as having humankind at the centre, or as the measure, or as the chief point of interest; to conceive of the gods as human beings writ large . . . to think that nothing has greater value than human beings, and that everything else can legitimately be bent to the service, use or interest of humanity, is to place humankind at the pinnacle of value in the world, and to privilege human existence over other kinds.

Grayling contrasts anthropocentrism with biocentrism, "the view that all life is valuable, not just human life" (27). This binary is central to the literature I discuss in subsequent chapters, though there are variants on, and degrees of, anthropocentrism, ranging from the notion that *Homo sapiens* ("wise species") is the only species that matters on the earth and that everything—the earth, animals, the oceans, and so on—belongs to it to exploit at will, to "softer" types of anthropocentrism that may be politically more practical than sheer ecocentrism or biocentrism. A soft or "weak" anthropocentrism states that while humans have more intrinsic value than nonhuman nature, the latter also has intrinsic value deserving respect up to the point of vital human needs.[4] Since humans appear to be uniquely self-conscious, it may be true that a completely nonanthropocentric view is finally impossible, but it would be disingenuous to conclude, as some have, that the matter is settled there. Mary Midgley makes a distinction between the concept in the personal sense of "self-centered," which is a good thing, accounting for a healthy sense of own's sense of balance in life, and its troubling application to our species in which people see themselves as "the absolute, objective center of everything" (103).

Contrary to anthropocentrism, ecocentrism (ecosystem-centered ethics) places intrinsic value on all forms of life, independent of their value for humans. Biocentrism, a closely related term, is a life-centered ethics which, writes Paul W. Taylor, accounts for "the obligations and responsibilities we have with respect to the wild animals and plants of the Earth." These responsibilities "arise from certain moral relations holding between ourselves and the natural world itself. The natural world is not there simply as an object to be exploited by us" (12). Rather, natural things possess inherent worth (13). As I show, the confronting of, and opposition to, anthropocentrism are central in the work of many fiction writers, poets, creative nonfiction writers, and literary essayists as well as environmental philosophers such as Taylor and Aldo Leopold, whose watershed *A Sand County Almanac* (1949) focuses on a "land ethic," which "changes the role of *Homo sapiens* from conqueror of the land-community to plain member and citizen. It implies respect for his fellow-members, and also respect for the community as such" (240). Rachel Carson's landmark book *Silent Spring* (1962), which shows the effects of human meddling with the natural world, set off a new wave of environmental consciousness and activism and inspired a number of environmental philosophers, including Norwegian philosopher Arne Naess, who coined the idea of "deep ecology" as a contrast to traditional environmentalism, which was, according to Naess, largely anthropocentric in its concerns. Since it provides the ideological basis by which industrial technology is employed to dominate nature, anthropocentrism is, for environmentalists, an attitude that must be rejected.

Val Plumwood, who in turn criticized deep ecology as androcentric, shows that anthropocentric criticism is central to environmental ethics;

"human-centeredness must be challenged in the politics of any serious ecological movement, whether this is done explicitly or not" (*Feminism* 141). Anthropocentrism leads to ecological denial and blindness; it "tends through incorporation to represent the Other of nature entirely in terms of human needs" and "encourages a massive denial of dependency" and a sense of "hyperseparation" (142). She discusses "anthrocentrism" within the context of hegemonic centrisms such as Eurocentrism and androcentrism and other major, entrenched cultural norms. Plumwood espouses a feminist, liberation model, which does not call for the impossible abandonment of a human epistemology but rather "an emphasis on human continuity with non-human nature" and "an understanding of human embededness in nature" ("Androcentrism" 139–140). Plumwood notes that philosophers seldom discuss anthropocentrism, and when they do, they routinely dismiss it as unfeasible or as an epistemic impossibility. She discusses and counters some of these arguments in her book *Environmental Culture* (see especially Chap. 6). One line of argument philosophers use revolves around what Plumwood calls a "false choice between self-abnegation or egocentrism—either totally neglecting or being totally enclosed in our own interests . . . Human-centredness is no more inescapable than any other form of centrism" (134). Holding a nonanthropocentric view "does not require us to eliminate either our own interest or our own locatedness, rooting out any trace of our own experience and any concern for our own needs" (132–133). Other environmental philosophers, including J. Baird Callicott and Holmes Rolston III, have argued forcefully in favor of a nonanthropocentric ethics.

Important studies by scholars such as Lynn White, Jr., Clarence J. Glacken, and Carolyn Merchant that delineate the development of human attitudes toward nature preclude the need for an extended survey of the history of anthropocentrism in the West.[5] The history of Western thought is filled with anthropocentric assumptions about the world. It is rooted, in part, in the Bible, beginning with its first chapter in Genesis, which many understand as a divine charge for absolute human sovereignty. After creating Adam and Eve, God commands them to "replenish the earth, and subdue it: and have dominion over the fish of the sea, and over the fowl of the air, and over every living thing that moveth upon the earth" (1:28). Echoing Lynn White, Jr., German theologian Jürgen Moltmann writes that an anthropocentric reading of this and other Bible verses drove out the cosmocentrism of the ancient world ("Christianity" 356). Historically, the biblical charge of ownership has legitimized a hard-line anthropocentrism, and it has been cited by everyone from Francis Bacon in his Great Instauration, which is modelled after the six-day creation in Genesis, to the modern-day Christian Dominionists, who see the virtual destruction of creation as a positive good.[6]

The Hebrew Wisdom books to some degree moderate the hard-line anthropocentric stance. Psalm 8 opens by invoking the image of "man" as a small, humble member in the chain of creation, but the remainder of the Psalm suggests what would later be known as humanism:

> When I consider thy heavens, the work of thy fingers, the moon and the stars, which thou hast ordained;
> What is man, that thou art mindful of him? and the son of man, that thou visitest him?
> For thou hast made him a little lower than the angels, and hast crowned him with glory and honour.
> Thou madest him to have dominion over the works of thy hands; thou hast put all things under his feet:
> All sheep and oxen, yea, and the beasts of the field;
> The fowl of the air, and the fish of the sea, and whatsoever passeth through the paths of the seas. (8:3–8)

Here, as elsewhere, the Bible appears to espouse an utter anthropocentrism —a human-centeredness, since it revolves around the idea of God's plan for humanity (for Christians, God himself having become a man)—but it is also filled with passages extolling the beauty of God's creation in itself. Psalm 148 is notable in part for its description of a natural world—mountains, valleys, trees, birds, wild goats, badgers, and so on—with very little mention of humans, suggesting that God created a world not only for people but for the beasts as well. The anthropocentrism in one of the foundational anthropocentric texts, the Old Testament, is not always as rigid as assumed.

Yahweh, in his long speech at the end of Job, asks a series of rhetorical questions. Suggesting that God provides rain for places (and presumably its creatures) outside of human use, he asks who caused "it to rain on the earth, where no man is; on the wilderness, wherein there is no man" (38: 26). Job 14 is a rather forthright nonanthropocentric view of world. Man's days are few, unlike a tree, which, when trimmed, sprouts again. When a human dies, that's it. As the celebrated translator Robert Alter has written, Job is a "radical rejection of the anthropocentric conception of creation that is expressed in biblical texts from Genesis onward" (3). In Chapter 40, Job replies meekly (in Hebrew Wisdom parallelism) to Yahweh, "I see how little I am,/I will not answer You/I am putting my hand to my lips." Yahweh goes on to underscore Job's confession of smallness by a heavily metaphorical and much-discussed evocation of the wonders of wild nature, including the behemoth, who is "the first of God's ways," and the leviathan.

Some Eastern religions, Daoism in particular, teach that humans are not the center of the world, that we must attune ourselves to the universe. The Daoist philosophers Lao Tzu and Chuang Tzu taught that each natural being has its own good and precious reality that preclude a human-centered world. In Confucianism, humans are understood as "anthropocosmic"—a microcosm in relation to the macrocosm of the universe (Tucker 410). Citing examples from Sri Lanka, Japan, and Thailand, Patrick Curry shows how Buddhist-inspired movements have fought against deforestation (146–147). Gary Snyder's *The Practice of the Wild* (1990) is a persuasive statement regarding an ecological ethic merged with Zen Buddhism. Many animistic religions of ancient cultures and indigenous societies hold what might be called anthropocentric viewpoints. But other religions, especially Western ones, promote human-centered views. Islam, historically, has held an anthropocentrism similar to that of Christianity. Allah made humans "successors upon the earth" (*Qu'ran* 6:165) and created the stars for human navigation (e.g., 2:29, 6:96). Turkish Sufi master Abdülhamit Çakmut says, "Everything is meant to serve man. If people are gone from this cycle, nature itself will be over" (qtd. in Weisman 270). Yet even in Islam the lines are not always so strictly drawn. Some Muslims, including Seyyed Hossein Nasr, cite *Qu'ran* 15:19—"And the earth—We have spread it and cast therein firmly set mountains and caused to grow therein [something] of every well-balanced thing"—as a point for an understanding, and acting on behalf, of ecological balance.[7]

Generally, the writings of the early church fathers, St. Augustine, and even *The Mirror of Perfection* (the anonymously written book about St. Francis) and Thomas Aquinas lean heavily toward anthropocentrism, and large threads of this thinking persist today. For example, Paragraph 2415 of the revised *Catechism of the Catholic Church* (1994) states, "Animals, like plants and inanimate things, are by nature destined for the common good of past, present and future humanity" (580). Yet there is also much praise of nature as a repository of God's creation in all these writers. St. Bonaventura waxes toward pantheism by stating, "The creatures of this sensible world signify the invisible things of God, partly because God is of all creation the origin, exemplar, and end . . ." (2.12). Some Christians connect *contemptus mundi* with the New Earth referenced in Revelation 21:1–7 and is both an effect and cause for a generally exploitive view of the natural world.

Addressing the relationship between religion and nature, Clarence J. Glacken writes,

> To say of the earth that it has been designed by a Creator for the sake of all life is one thing: to say that it is made for man alone and to use as he sees fit is another. The anthropocentrism of the latter is narrow and crippling. In the secular versions, which follow, the narrow and crippling anthropocentrism continues in the assumption of universal utility for man. All nature becomes a resource. ("Man" 133)

Yet the notion that Judaism and Christianity are the sole bases for anthropocentrism in the West does not bear close scrutiny. Philosophically, the notion of a human-centered universe is rooted in sources well before the rise of Christianity and outside Jewish traditions. The first and greatest sophist, Protagoras, stated famously that "Man is the measure of all things" (Waterfield 211), yet Plato quotes Protagoras' famous dictum in his *Theaetetus* (152a) less to proclaim anthropocentrism than to show that truth is relative for each person (see Guthrie 67–68). The writings of Plato and especially Aristotle are heavily anthropocentric. My second chapter addresses the development of the idea in their writings but more centrally the idea's subversion by other ancient Greeks and Romans.

In *Antigone*, Sophocles has much to say about the glories of man, as well as his failings. Through his inventiveness, man is "lord . . . of the beasts" (348–349).

> And speech he has learned, and thought
> So swift, and the temper of mind
> To dwell within cities, and not to lie bare
> Amid the keen, biting frosts
> Or cower beneath pelting rain;
> Full of resource against all that comes to him
> Is Man . . . (352-61)

Echoing Sophocles, Hamlet's famous speech on "man," addressed to his friends Rosencrantz and Guildenstern, similarly underscores the miracle of the species, though the speech finally underscores Hamlet's pessimism and conviction that we are little more than dust in an impressive natural backdrop:

> . . . this goodly frame, the earth, seems to me a sterile promontory. This most excellent canopy the air, look you, this brave o'erhanging, this majestical roof fretted with golden fire—why, it appears no other thing to me than a foul and pestilent congregation of vapours. What a piece of work is a man! How noble in reason, how infinite in faculty, in form and moving how express and admirable, in action how like an angel, in apprehension how like a god–the beauty of the world, the paragon of animals! And yet to me what is this quintessence of dust? Man delights not me—no, nor woman neither, though by your smiling you seem to say so. (Shakespeare 2.2.301-12)

The writings of Shakespeare's contemporary Francis Bacon are an influential anthropocentric source. Like the Bible, the English Authorized Version of which appeared during his mature years, his writings reflect both a respect for nature on the one hand but the charge to exploit it on the other. Man is, he writes, "the servant and interpreter of nature," and that "Nature to be commanded must be obeyed" (39). Yet he establishes an ontological divide by stating that man is outside nature. Carolyn Merchant pinpoints the legal, patriarchal language in many of Bacon's works (e.g., "penetrate into the inner and further recesses of nature"), but she notes that Bacon likely did not intend for his words to justify the wanton destruction of the natural world. Undoubtedly, they nonetheless contributed heavily to that end (see Merchant, Chap. 7). Naomi Oreskes and Erik M. Conway, in their speculative but science-based account of our environmental crisis, write that Baconianism would have dire consequences for the planet in its failure to control climate change: technological and economic power "did not reside in the hands of those who understood the climate system, but rather in political, economic, and social institutions that had a strong interest in maintaining the use of fossil fuels"—a strong system conglomeration of energy interests the writers predict that future historians will look back on and call "the carbon-combustion complex" (36). Descartes, who espoused the separation of humans from the rest of the natural world, is another important anthropocentric source. Dividing humans from the rest of the natural world, he refers to (nonhuman) animals as "automata" (359), an idea that bore strong influence and wide contestation.[8]

Immanuel Kant writes in *The Critique of Judgement* (1790), "As the single being upon earth that possesses understanding, and, consequently, a capacity for setting before himself ends of his deliberate choice, [man] is certainly titular lord of nature, and, supposing we regard nature as a teleological system, he is born to be its ultimate end" (2. 93–94).[9] Yet he follows David Hume (*Dialogues Concerning Natural Religion*) in expressing doubts about an anthropomorphic God. Nature is not designed for us, but, Kant states, we must conceive of nature with a moral anthropocentrism as a sort of heuristic aid. This, writes Paul Guyer, "leads us to the view that we must think of nature as if it were meant to be an arena hospitable to the realization of the goals of human morality" (338). Similarly, although Kant rejects machine and artisan analogies in representing nature, an analogy "may be used by reflective judgement as a regulative conception for guiding our investigation" and reflection (2.23–24). Written around the time of Kant's *Critique of Judgement*, James Hutton's *Theory of the Earth* (1788) attempts to reconcile the Bible and science, with the result of an utterly anthropocentric text. Hutton states that the world resembles a machine or a body and that "the system of this earth has either been intentionally made imperfect, or has not been the work of infinite power and wisdom." The earth, he continues, "is evidently made for man," and only man can know the nature of the world (216–17).

Anthropocentric statements as well as antianthropocentric ones come from many directions, sometimes unexpected. Less than a decade after publishing a groundbreaking study with Charles Darwin on natural selection in 1858, Alfred Russel Wallace embraced spiritualism and began participating in séances. Concluding his book *Darwinism*, Wallace wrote (to Darwin's dismay) that "the whole purpose, the only *raison d'être* of the world—with all its complexities of physical structure, with its grand geological progress, the slow evolution of the vegetable and animal kingdoms, and the ultimate appearance of man—was the development of the human spirit in association with the human body." During the first two decades of the twentieth century, Gifford Pinchot, the first head of the United States Forest Service, appealed to anthropocentric reasoning in support of damming the Hetch Hetchy Valley (a victory he would win), while naturalist and writer John Muir fought for its preservation on ecocentric (or perhaps ecotheocentric) principles (see Nash Chap. 10). Written around the time of the Hetch Hetchy controversy, Charlotte Perkins Gilman's female utopia novel *Herland* (1915) has merit in envisioning the full, realized personhood of women outside of patriarchal terms. But its carefully planned, sanitized society is anthropocentric, anticipating the dystopias of Zamyatin, Huxley, and Orwell: forests have been razed and replanted with nut and fruit trees that have human use, and all species of animals not immediately useful to people have been exterminated. No "wild beasts" and little biodiversity remain. The novel's narrator, Van Jennings, remarks indignantly but erroneously to his love interest regarding his homeland, "We are not animals!" (138).

The predator-cattle controversy in the American West remains divided, often between these two poles. For many of the owners and residents, the land exists *for* human exploitation, while for others, wolves, bears, and other predators have a longstanding claim to existence and need of a large, free range for survival. The wholesale killing of predators would have enormous effects for the ecological balance in Western United States. In *The Ninemile Wolves*, Rick Bass, writing within the context of anti-wolf activities in the Southwest, cites an anti-wolf film that states, "We can go Man's way, or Nature's way" (85), which suggests erroneously that humans have a "way" that is not "natural" and that to not go "nature's way" is ever desirable. Later in the book, Bass notes the talk of two "old boys" in a bar who talk of "deep-sixing" the wolves of Montana; this, writes Bass, would be "deep-sixing culture" (117). The notion that environmental matters must be "balanced" with economic ones is based on the false assumption that environmental concerns are a mere luxury. Good ecological practice demands the diversity of species. Jared Diamond writes that the elimination of a "minor" species compares with randomly knocking out rivets that hold an airplane together (*Collapse* 503, 489).

Subsequent chapters show that a nonhuman centered view is *not* a recent development, yet the influence of this thought on our actions and attitudes has been light at best. As James Gustave Speth writes, "The environmental community has grown in strength and sophistication, but the environment has continued to deteriorate" (x). We act anthropocentrically by choice, against the grain of ecological science. Twentieth-century United States environmental policies indicate small movements away from anthropocentric bases of thought. American environmentalists in the early part of the century were opposed strongly to automobiles in national parks and other wilderness areas. Ecological science "pushed preservationists beyond scenery to embrace a purer wilderness ideal"—a more sophisticated tool for opposing the anthropocentrism of utilitarian conservationists and aesthetic limitations of park preservation (Sutter 168). The influence of ecological thinking on the birth of modern wilderness policy was small, yet it grew in impact with Leopold and other wilderness advocates by the 1940s (Sutter 177; M. Lewis 206). Edward Abbey's "Polemic: Industrial Tourism in the National Parks" in *Desert Solitaire*, which argues that the building of roads in parks violates the Wilderness Act, bore influence on the Sierra Club and the Conservation Foundation, as well as more radical environmental groups (Wild 196; Little 56).

However, there is not a lot of tangible evidence that nonanthropocentric thought has affected how we have treated the earth and our fellow species in public policies and in practical affairs, even though (the central idea in my following chapters) anthropocentrism has been contested for millennia. Opposition to anthropocentrism began in antiquity, in Greek and Roman writings, at the latest, though the idea arose earlier in Eastern philosophies and religions. The critique of anthropocentrism is not attached to any one philosophy or religion. It was developed among Cynics, Stoics, and Epicureans; it is pronounced by theists, atheists, agnostics, and pagans. It is a point of view that extends from antiquity, to the early Christian era through the Middle Ages, the Renaissance, the Enlightenment, into the Romantic era and into the twentieth and twenty-first centuries. It appears in cases as a romantic idea, perhaps, but it is also a feature of writers who are opposed diametrically to romanticism. By the later nineteenth century it was a positivist commonplace, confirmed by modern biological and other sciences.

Despite its long history, anthropocentrism would not be questioned prominently in science until the seventeenth century with the work of Galileo. Galileo's conception of motion, writes Charles Coulston Gillispie, "altered man's consciousness of a real world outside himself in nature" (42). Regarding the impact of the writings of Galileo (and the stage being set for Cartesian dualism), E.A. Burtt writes, "Man begins to appear for the first time in the history of thought as an irrelevant spectator and insignificant effect of the great mathematical system which is the substance of reality" (90). Albert Einstein praises Galileo's gifts as a writer to "overcome the anthropocentric and mythical thinking of his contemporaries" and lead thinkers in his day to a scientific view of the cosmos (vii). Part of Galileo's argument in his 1615 *Letter to the Grand Duchess Christina* is against the anthropomorphic view of God as a beginning point in overturning the hard anthropocentrism of the day. Then he proceeds to argue that the Bible is not a valid standard for science but a spiritual book that should sometimes be read allegorically, lest one be misled by its obvious mistakes. A devout Christian, Galileo quotes not only the Bible but the Church Fathers, including Tertullian and Augustine. Nature, he says, is a book that we must read in order to understand God's creation. Elsewhere, Galileo posits that the old Aristotelian science has had its day. In his *Dialogue Concerning the Two Chief World Systems*, written at the age of sixty while imprisoned, he states, "I say that it is brash for our feebleness to attempt to judge the reason for God's actions, and to call everything in the universe vain and superfluous which does not serve us" (368)—a startlingly antianthropocentric statement and one of the first in the early modern (scientific) era.

Working from ideas established by Copernicus (who drew heavily from Ptolemy), Galileo lay the foundation for a scientific overturning of anthropocentrism, but Darwin marks its most decisive turning point. Although frameworks for the interrelations in the organic world existed before Darwin among the physico-theologians, Darwin established them as a fact in scientific, evolutionary terms (see Glacken, "Man" 134). As early as 1838, twenty-one years before publishing *On the Origin of Species*, Darwin wrote in notebook, "Man in his arrogance thinks himself a great work, worthy the interposition of a deity. More humble and I believe true to consider him created from animals" (*Notebook C*). His *On the Origin of Species* (1859) would argue subtly that humans and all other species are interrelated organically, the result of a common descent. Near the conclusion of the book, he would write, "When I view all beings not as special creations, but as the lineal descendants of some few beings which lived long before the first bed of the Silurian system was deposited, they seem to me to become ennobled" (395). His critique of anthropocentrism with regard to the special creation of humans is implicit in *On the Origin of Species*; *The Descent of Man* (1871) shows more explicitly that humans and apes were constructed from a common model. It was, he writes, the "natural prejudice" and "arrogance" of our forefathers that humans are the product of "demi-gods" (25). Darwin knew little about genetics, but the Human Genome Project, completed in 2003, showed that humans and chimpanzees are nearly identical in their genetic composition ("New Genome").

The examination of anthropocentrism as well as ecocentrism, biocentrism, and its variants, including the questioning of anthropocentrism with or without explicit ecological ideas, is central to ecocriticism. The works I discuss concern the relationship between the human and nonhuman, and this, writes Laurence Coupe, is "the most fundamental question of all for green studies" (119). A growing number of disciplines inside and outside literary and rhetorical studies show the centrality of overcoming anthropocentrism. Writing about postcolonial ecology, Elizabeth DeLoughrey and George B. Handley posit that "the ecocritical interrogation of anthropocentrism offers the persistent reminder that human and political and social inequities cannot be successfully and sustainably resolved without some engagement with the more-than-human world" (25). Whereas anthropocentrism is inherent in Gestalt therapy, Andy Fisher writes that the field of ecopsychology by contrast seeks to negate anthropocentrism (102). Tim Ingold proposes an ecology "in which organism and environment are understood as internally related, the whole-organism-in-its-environment being an indivisible developmental system, 'the creative unfolding of an entire field of relations'" (qtd. in Fisher 92–93). Hard sciences such as geology have potentially transformative powers and the ability to teach humility:

> the geologic sciences disclose the measure of the gods. That is, the geologic sciences provide a nonhuman standpoint for us to view life and experience, helping us look past the interests and perspective of the small self to the larger self that is part of nature. And when they do so, these sciences become an imaginative engagement with the world, an engagement that leads to a profound appreciation for the uniqueness and preciousness of the life of the planet—carrying us, in the end, toward an abiding compassion for all creatures bound up in time's arrow. (McLean, Moores, and Robertson 239)

Working from an animal rights/environmental ethics perspective, Gary Steiner's overview of the moral status of animals in Western philosophy traces thought from antiquity (e.g., Xenophanes) through Kant, Darwin, Heidegger, and into the postmodern era that seeks just conceptions of the human-animal relationship. Although discussion of animal rights accelerated beginning in the twentieth century, "the leading contemporary theories of the moral status of animals," he writes, "ultimately privilege the interests of human beings over nonhuman animals" (4).

This book is not about economics, yet the elephant in the room in a discussion of anthropocentrism centers on the relationship between capitalism and our attitudes and actions toward the planet. Garrett Hardin's oft-cited 1968 essay "The Tragedy of the Commons" counters the classic argument that growth is an economic good in itself. Growth, Hardin states, leads to exploitation and annihilation. It would be shortsighted to state that capitalists are by nature anthropocentric (some may espouse a capitalism checked through regulation) or that only capitalists are responsible for our anthropocentric views. The cases of China and the former USSR show that capitalism is not the only system that ravages natural environments. Perhaps unchecked industrialization is the most responsible perpetrator of anthropocentrism en masse, though China began embracing global capitalism in the late twentieth century, as did Russia. Echoing other critics, Slavoj Žižek writes that the solution to our ecological crisis lies in the substantial tempering of capitalism. The 2009 Copenhagen conference held by twenty world powers to combat global warming resulted merely in vague statements with no action: "the state political elites serve capital, [and] they are unable and/or unwilling to control capital even when the very survival of the human race is ultimately at stake. Frederic Jameson's old quip holds today more than ever: it is easier to imagine a total catastrophe which ends all life on earth than it is to imagine a real change in capitalist relations . . . " (334).

In his proposal for making a sustainable market system that takes into account the environment, James Gustave Speth posits that capitalism is biased in favor of the present at the cost of the future and the environment. Our values today, he writes, "are strongly materialistic, anthropocentric, and contempocentric . . . . The anthropocentric view that nature belongs to us rather than we to nature eases the exploitation of the natural world" (62). James Atlas also reflects on the "contempocentrism" of our thought and actions: "History is a series of random events organized in a seemingly sensible order. We experience it as chronology, with ourselves as the end point—not the end point, but as the culmination of events that leads to the very moment in which we happen to live." But there is really no end point, no culmination. As a result of shortsighted, profit-motivated actions, wilderness and natural environments become pawns in a game played by a handful of venture capitalists, while humans and other living things suffer the consequences. If humans are in fact the cleverest animal on the planet, this does not justify the boisterous imprinting of our species on all the others and on our common earth. On occasion, it appears that Mary Shelley's vision of Frankenstein's monster ravaging the countryside once he is rejected by humans, or the title character of *Orlando Furioso* destroying everything in his path during his romp through Europe and Africa, are realistic pictures of our species.[10]

A common, and sometimes valid, reaction to environmentalism revolves around choices regarding human livelihood versus the safeguarding of other species and the protection of wilderness or environmentally sensitive land areas. Yet the choice is often based on a false dilemma. The protection of the environment and the protection of underclasses in particular are coterminous; without environmental justice, one cannot expect a lot of social justice. Appalachia residents living where coal industries have practiced mountaintop removal coal mining can speak for this, as can those living in areas most affected by unhealthy and toxic fumes produced by industries. Garbage dumps are simply not established in wealthy neighborhoods; they are almost always in financially stressed areas. Minorities are especially victims of such extreme anthropocentric practices.

Worldwide, political trouble spots and environmental trouble spots are very close to one and the same (see Diamond, *Collapse* 497, 516). Donald Worster finds a strong link between democracy and wilderness preservation. It is no accident that countries protecting wilderness tend to exercise democracy more fully than those that do not, though a poor country such as Costa Rico protects 28% of its land as wilderness, whereas the United States protects only 4.7% of its lands (266). The idea that humans have dominion over nature—probably the dominant view among current elected officials in the United States—is not sustainable. The natural world is a partner with democracy. Worster writes that a preservation ethic is a threat to tyrants, to authoritarian minds such as Stalin, Mao, and Pol Pot (269). The biggest challenge to conservationists, he writes, lies in the reality that nondemocratic countries "that allow no dissent from orthodoxy, no openness to new ideas or research, and no respect for the other-than-human world" control the destinies of many of the world's remaining wilderness areas (270). Although perceived by some, erroneously, as a white, middle-class issue, wilderness protection is in itself a social issue, inextricably connected to human freedom. Howard Zahniser, whose bureaucratic work in Washington was vital in establishing and passing the Wilderness Act of 1964, said in a speech, "To know the wilderness is to know a profound humility, to recognize one's littleness, to sense dependence and interdependence, indebtedness, and responsibility" (qtd. in Harvey 193). Much of the power of the writings of William Wordsworth, Henry David Thoreau, John Muir, and Edward Abbey, among others, lies in the discussion of the vital connection between wilderness and freedom.

Posthumanisms

My present interest is in ways shared by those pursuing posthumanist studies, a field that does not present a unified view. Rosi Braidotti writes that "the humanist image of thought . . . sets the frame for a self-congratulating relationship of Man to himself" and that it is "crucial for Humanities and for critical theory to find adequate language for a postanthropocentric world" (67, 82). This is surely true, but, as I show, writers posited firmly in the humanist tradition have challenged and subverted anthropocentrism and have already provided some of the language needed to confront and perhaps overcome extreme forms of anthropocentrism. I am interested in the "posthuman" not in its application to technology or cyborg studies (e.g., Bostrom, Kurzweil and Hayles), but perhaps more eccentrically, from a literary-historical perspective that functions rhetorically, as a large body of work that seeks an ecological understanding of the place of humans in the world. Attitudinal changes that seek a drawing down of the Anthropocene might result from this understanding.

The term "posthuman" has been applied in many different ways since the early twentieth century. H.P. Lovecraft employs the term ("post-human") twice in his 1936 story "The Shadow out of Time." Neil Badmington notes that H.P. Blavatsky used the term *"post-Human"* in 1888 ("Posthumanism" 376). The modern use of the term is rooted in the writings of Donna Haraway, though her seminal essay "A Manifesto for Cyborgs" (1985) does not use the term or its variants. Although the works I address are generally removed from cyborg studies, the overlappings are impossible to account for. Yet my work is tied to Haraway's concept of nature-cultures which, write Ivan Callus and Stefan Herbrecht, "require new (posthumanist) forms of ecocriticism that challenge views of human 'dominion' over the world and instead acknowledge the multitude of interactions and mutual interdependencies between humans, nonhumans and their environment" (149).

It is difficult to locate the philosophical (distinct from the ecological) roots of posthumanism or antianthropocentrism, terms that are, in my usage, nearly synonymous.[11] Some of those roots lie in the works of some of the ancients I discuss in the next chapter. Others lie in the writings of Baruch Spinoza, who argues in *The Ethics* that the belief in final causes in nature is based on ignorance and are "mere human figments" (77). In opposition to Cartesian and Newtonian materialism, he posits that the understanding of our species' relationship with the rest of nature is central to human meaning and happiness (Oelschlaeger 121–22). Some of the roots lie with Arthur Schopenhauer, whose pessimism is based on a conception of the world and human nature as wretched—in his philosophy a necessary truth based on the underlying reality of the Will. Friedrich Nietzsche marks the starting point for the rejection of humanism that would develop in the twentieth century, even though what we may now call posthuman ideas existed well before Jacob Burckhardt (Nietzsche's older colleague at the University of Basel) and his groundbreaking historiographies on the Renaissance in 1867.[12] Nietzsche writes in *Human, All too Human* that philosophers "involuntarily think of 'man' as an *aeterna veritas* [something everlastingly true], as something that remains constant in the midst of all flux, as a sure measure of things." Philosophers lack historical sense, assuming that humans in the here and now comprise humans for all time. "But everything has become: there are *no eternal facts*, just as there are no eternal truths" (12–13).

The key to Nietzsche's antihumanism (and his antianthropocentrism) is what he termed his "extravagant task" to "translate man back into nature" (*Beyond* 162), to return humans to their naturalistic setting, free of the religious agency and conventional morality to which his writings are openly hostile. In context, it would, I think, be difficult to read Nietzsche as anything like an environmental philosopher, yet his viewpoint may easily be mistaken as biocentric if one reads much of his work out of context. In *The Antichrist*, for example, he writes,

> We no longer derive from "the spirit" or "the deity"; we have placed him back among the animals . . . Man is by no means the crown of creation: every living being stands beside him on the same level of perfection. And even this is saying too much: relatively speaking, man is the most bungled of all the animals, the sickliest, and not one has strayed more dangerously from its instincts. (580)

But he goes on to state that man is, "of course, the most interesting" animal. Nietzsche believed that humans are exceptional, that dignity is a necessary illusion for human survival, and that people always have and always will dominate the planet. Environmental philosopher Michael Zimmerman writes that Nietzsche would not have thought well of E.O. Wilson's concept of biophilia, the idea of an instinctive bond between humans and other living things ("Nietzsche" 14–15, 17). Yet Nietzsche's decentering of human subjectivity has had a strong impact on posthumanism, animal studies, ecocriticism, and other fields that call anthropocentrism into question.

The contributions of Martin Heidegger to posthumanist studies are at least as important as Nietzsche's, and they are at least as problematic to appropriate for environmental ethics. His later work is critical of the Western tradition of technology as "enframing" (*Gestell*), nature as a standing reserve on call for human purposes. Aligning Heidegger with deep ecology, Zimmerman writes that for Heidegger, "Western metaphysics led not to human 'progress,' but instead to technological nihilism in which everything—including humankind—stands revealed as raw material for the goal of greater power and security. According to Heidegger, this arrogant humanism (whether capitalist or communist) not only diminishes humankind, but also wreaks havoc on nature." (These ideas recall Nietzsche, who writes in *On the Genealogy of Morals*, "Our whole attitude toward nature, the way we violate her with the aid of machines and the heedless inventiveness of our technicians and engineers, is *hubris*" (549).) Zimmerman aligns such thinking with that of deep ecologists: "this crisis is not accidental, but instead is a symptom of the arrogance of anthropocentric humanism, which diminishes humankind while wantonly destroying nature" ("Heidegger" 3–4). Zimmerman would later revise his ideas on the relationship between Heidegger and deep ecology on the basis of Heidegger's National Socialism: Heidegger's thinking and deep ecology "are in many ways incompatible, in part because deep ecologists—in spite of their criticism of the ecologically destructive character of technological modernity—generally support a 'progressive' idea of human evolution" ("Rethinking" 1).

Heidegger was a nature lover and a critic of anthropocentrism, but, as Zimmerman realizes, his alignment with environmental philosophy is problematic. Tony Davies writes that Heidegger "strove in his critique of metaphysical and rhetorical error to position himself outside the assumptions of European thinking since Plato, and the anthropocentric illusion that lies at its heart, insisting that 'Man' is not the imperious subject but merely the object, the recipient, of 'Being', not the creator of language but its creature" (129). In his "Letter on Humanism" (1947), Heidegger writes that *Desein* ("being there") is most realized in humans, that "Man is the shepherd of Being" (234). For him, "Humanism is a conception of man according to which man occupies a central place (*eine Mitte*) within the totality of beings" (Philipse 199). Yet Heidegger insists that this idea does not warrant an anthropocentric response. For the postwar Heidegger, traditional conceptions of humanism (Christian, Marxist, etc.) presuppose a metaphysical stance. Hence humanism is inadequate to the high essence of man (the guardian of Being who lives in language): "man only resides in his proper essence, if he is claimed by Being . . . Heidegger interprets the history of metaphysics as a history of abandonment by Being" (Philipse 199–200). James Luchte writes, "Being and Time is opposed to 'humanism'. Yet, this higher essence of 'man' is not meant in the sense of metaphysical subjectivism (or as anthropocentrism) in which man is the tyrant of Being to which each and all is subject."[13]

Following Nietzsche's critique of humanism, in *The Order of Things*, Michel Foucault addresses humans, or the human subject, as *episteme*, as the center of discursive formation.[14] The concept of "humans," in the Burckhardt sense, is a new invention. He concludes his book with a passage that is among the most often quoted in the writer's oeuvre:

> As the archeology of our thought easily shows, man is an invention of recent date. And one perhaps nearing its end.
>
> If those arrangements were to disappear as they appeared, if some event of which we can at the moment do no more than sense the possibility—without knowing either what its form will be or what it promises—were to cause them to crumble, as the ground of Classical thought did, at the end of the eighteenth century, then one can certainly wager that man would be erased, like a face drawn in sand at the edge of the sea. (387)

Similarly, Bruno Latour writes in *We Have Never Been Modern*, "The human is not a constitutional pole to be opposed to that of the non-human. The two expressions 'humans' and 'nonhumans' are belated results that no longer suffice to designate the other dimension" (137). And more pointedly, Jean Baudrillard, echoing numerous works of fiction and poetry, writes, "One has only to consider the human being himself, complete with his emotions, his passions, his laughter, his sex and his secretions, to conclude that man is nothing but a dirty little germ—an irrational virus marring a universe of transparency" (34).

Many attacks on humanism in the twentieth century have come from outside the Nietzschean tradition. Louis Althusser writes that since humanism is itself an ideology, a precondition for Marxist philosophy "is theoretical anti-humanism" (33). Yet, for Marx, the demotion of humanism is not balanced by a promotion of nature; capitalism rightly rejected "the deification of nature": "thus nature becomes for the first time simply an object for mankind, purely a matter of utility" (qtd. in Midgley 104–05.) Far from Marx (and Althusser), T.E. Hulme, a Tory and proponent of a new classicism, writes in his 1924 essay "Modern Art and Its Philosophy" that the rise of modern science marks not the death but the rise of anthropocentrism: "The change which Copernicus is supposed to have brought about is the exact contrary of the fact. Before Copernicus, man was not the centre of the world; after Copernicus he was." Rousseau and the rise of romanticism were the high points for such thought (80). In a 1988 essay, Wendell Berry addresses the philosophical but more particularly poetic move away from the theme of nature as the great human instructor in English poetry. Berry cites the idea developed in the book of Job, Virgil, Spenser, Shakespeare, and Milton and, echoing Hulme, marks the eighteenth century as the starting point for the rise in anthropocentrism:

> After Pope, so far as I know, this theme departs from English poetry. Later poets were inclined to see nature and humankind as radically divided and were no longer much interested in the issues of a *practical* harmony between the land and its human inhabitants. The romantic poets, who subscribed to the modern doctrine of the preeminence of the human mind, tended to look upon nature not as anything they might ever have practical dealings with, but as a reservoir of symbols. ("A Practical" 105)

Later chapters will show that there is much truth in these statements, against the grain of science, yet the picture is not quite as simple as Hulme or Berry suggest. Rousseau and the romantics in his wake have much to say about anthropocentrism, often questioning and rejecting it outright. As Clarence J. Glacken writes, "one phase of the romantic movement" is "a rebellion against the dichotomy between man and nature. In the movement for the protection of nature, we own much to romantic ideas of the beauties of untouched and remote wildernesses, and of the importance of being conscious of our attitudes to the natural world" ("Man" 132).

Like any other strain of thought, antihumanism includes relatively optimistic or melioristic writers as well as decidedly pessimistic ones. In the latter camp are British philosopher John Gray and American horror fiction writer Thomas Ligotti. Commenting on Gray's ethical antihumanism, Tony Davies writes that "humanism is a form of collective narcissism, blind to its own folly, absurdity and cruelty" (147). Gray writes that humanism is a religion invented by Henri de Saint-Simon and Auguste Comte, and, a faith substituted for Christianity in the nineteenth century, is as much a sham as any organized religion (xii). Gray cites Schopenhauer as the first and unsurpassed critic of humanism (38). The fact that humans are uniquely conscious is, for Gray, not a big deal (59), and he quotes G.C. Lichtenberg: "That man is the noblest creature may be inferred from the fact that no other creature has contested this claim" (85).

Ligotti's nonfiction book *The Conspiracy against the Human Race* (2010) fills out the author's pessimistic philosophy in the tradition of Schopenhauer, fellow horror writer H.P. Lovecraft, and Norwegian philosopher Peter Wessel Zapffe, who saw human existence as a tragedy. For Ligotti, humans are not really a part of nature; we are dominated almost wholly by illusions. Only a small part of us is placed in the natural environment, and environmentalism "snubs the real issue" of facing and undoing the illusion that our lives matter in any real way (78). Humans are unique only by being conscious, a quirk of evolution:

> No other life forms know they are alive, and neither do they know they will die. This is our curse alone. Without this hex upon our heads, we would never have withdrawn as far as we have from the natural—so far and for such a time that it is a relief to say what we have been trying with our all not to say: *We have long since been denizens of the natural world*. Everywhere around us are natural habitats, but within us is the shiver of startling and dreadful things. Simply put: *We are not from here*. If we vanished tomorrow, no organism on this planet would miss us. Nothing in nature needs us. (221)

Ligotti is unsurprisingly skeptical of transhumanism as a viable answer to the human condition. It "encapsulates a long-lived error among the headliners of science: In a world without a destination, we cannot even break ground on our Tower of Babel, and no amount of rush and hurry on our part will change that" (127). Ligotti's stories, like Lovecraft's, work as laboratories for the author's mechanical philosophy, often with powerful results.[15]

For all the antihumanisms over the past century and a half, most of them, Kate Soper contends, harbor a humanist rhetoric (182), though, as later chapters show, the reverse is also true: many of the writers widely considered humanists harbor antihumanist rhetorics or at least make antianthropocentric statements. There is, to say the least, a large difference between current conceptions of "transhumanism" and the term as used by Robinson Jeffers in the 1940s (see Chap. 6). One difference between my work and that of posthumanist scholars is the fact that most of the latter are based thoroughly in an anthropocentric mindset: they are concerned with changing bodies, the nexus between human and machine, and this is very worthy, important work that will without a doubt be realized more fully in our increasingly technologized world. My focus is more set on geological time, the long haul, where humans may be seen as a blip on a screen. I do not, however, suggest that just because we are transient we are irrelevant: we are what we are, and, like all other species, we must struggle for existence and to find meaning in our lives on personal, philosophical, spiritual, and ethical levels. I do not seek out an essentially antihumanist philosophy, though the general direction may suggest otherwise; on the contrary, most of the literature discussed in the following chapters seeks to understand more fully our place in the world by questioning the notion of what some environmental philosophers have termed a "hard" anthropocentrism–the view that everything in the world exists solely for human benefit.

The Anthropocene

We live now in what many scientists have termed "the sixth mass extinction," a period of anthropogenic loss of species on a scale that would rank in geological terms as a catastrophe (Hallam 224). Over millennia, catastrophic events on the earth have caused major shifts in plant and animal populations. Scientists believe that five great shifts have occurred in the earth's distant past. Our species, which appeared somewhere around 200,000 years ago, has proved unique in its tendency to cause shifts at a greatly increased speed.[16] As a consequence, plant and animal species are becoming extinct at least 1000 times faster than before humans arrived (Pimm, et al.). According to E.O. Wilson, half of the animal species could be extinct by the end of the twenty-first century (see Conniff). Scientist-novelist David Brin delineates some of the specific ways in which humans have mismanaged the world, which he states is on the brink of ecological collapse:

> For carelessly cutting down forests and spilling garbage in the sea. For poisoning aquifers and ruining habitats. For changing the very air we breathe. For causing temperatures to soar, glaciers to melt, seas to rise and deserts to spread. For letting the planet's web of life get winnowed down, through biodiversity loss, till it's a fragile lattice, torn by any breeze. Most animals have the sense not to foul their own nests.

Humans have long had the ability to wipe out a massive number of species. Although the idea is controversial, some researchers state that about 11,000 years ago, humans (Paleoindians) were largely responsible for the Pleistocene extinctions. This theory is based on findings that the extinctions occurred after the appearance of humans, that only large mammals (sources of food and clothing) became extinct, and that no significant climate disaster appeared during this period (Hallam 225; Diamond "The Present"). Much more recently, in *Of Plymouth Plantation*, William Bradford writes that in 1621 the English colonists made it through winter with a store of fowl "of which this place did abound when [the colonists] came first (but afterward decreased by degrees)" (100). Writing in 1864, George Perkins Marsh notes, "There are parts of Asia Minor, of North Africa, of Greece, and even of Alpine Europe, where the operation of causes set in action by man has brought the face of the Earth to a desolation almost as complete as that of the Moon . . . The Earth is fast becoming an unfit home for its noblest inhabitant" (138). Today, "our activities are currently wiping out between 3000 and 30,000 species a year, from an estimated total of just 10 million" (McGuire 170).

The human effect on the planet has been so extreme that Nobel-Prize winning chemist Paul J. Crutzen sought in a well-known *Nature* article to establish the renaming of our era, supplanting what is commonly identified as part of the Holocene, the "Anthropocene":

> For the past three centuries, the effects of humans on the global environment have escalated. Because of these anthropogenic emissions of carbon dioxide, global climate may depart significantly from natural behaviour for many millennia to come. It seems appropriate to assign the term "Anthropocene" to the present, in many ways human-dominated, geological epoch, supplementing the Holocene—the warm period of the past 10–12 millennia. (23)[17]

Crutzen concludes his piece by stating that it is the task of engineers and scientists "to guide society towards environmentally sustainable management during the era of the Anthropocene. This will require appropriate human behaviour at all scales" (23). Although humans have a very long history of altering natural environments, modern technology and global capitalism have greatly increased this capacity. A recent report shows that over two decades, from the early 1990s to the early 2010s, humans have eliminated 10% of the world's remaining wilderness areas, the Amazon being the hardest hit area ("A tenth").

The American public in general and many politicians and industrial leaders in particular have not been willing to listen to the consensus of scientists when it comes to climate change. In a *New York Times* column, philosopher Gary Gutting states that "nonexperts," which means almost all of us, "are in no position to argue against the consensus of scientific experts." Yet the empirical evidence for climate change is overwhelming. Glacier National Park in Montana had over 150 glaciers in the 1880s, but as of the early 2000s only thirty-three glaciers remained. At the current rate of melting, soon none will be left (Diamond, *Collapse* 50). Glaciers in the Alps, Rockies, and Himalayas are also retreating rapidly ("Speed"). Caused largely by increased carbon dioxide and other greenhouse gases that humans have emitted into the atmosphere, nearly every successive year so far in the early twenty-first century has been the hottest on record.[18]

The media has not done its job well enough as the watchdog citizens required for democracy to work well. The failings of the media have something to do with the bottom line demand for profits: anthropogenic global climate change is simply not sexy, in the way that scandalized celebrity behavior is. Worse, the corporate media fears that reports on anthropogenic catastrophes will turn off a public that does not watch TV or read news in order to feel guilty but to be entertained. Consequently, the corporatization of the media has played a giant role in the public's lack of awareness regarding our negative impact on the planet. Ratings and bottom lines govern the coverage of TV and much other news media today. People don't want to believe that humans have trashed the planet as badly as the majority of scientists state, but the fact is that we have trashed it even worse than many scientists have predicated. As following chapters show, art and literature have substantial roles in changing behavior and changing attitudes, the latter being, writes Kenneth Burke, an "incipient act" (42). Although few of the writers I discuss in subsequent chapters approach directly the scientific analyses I cite, the human-centered assumptions that are largely the cause for our global crises are precisely the terrain of much of their work.

Biologist E.O. Wilson has suggested the "Eremozoic" as the term to account for the current era, or that of the very near future:

> The human hammer having fallen, the sixth mass extinction has begun. This spasm of permanent loss is expected, if it is not abated, to reach the end-of-Mesozoic level by the end of the century. We will then enter what poets and scientists alike may choose to call the Eremozoic Era—the Age Loneliness. We will have done it all on our own, and conscious of what was happening. God's will is not to blame. (91)

An extreme anthropocentrism lies at root in our mistreatment of the earth in the assumption, largely unexamined, that it exists for us to exploit; it may be that this assumption is a minority view, but it is seemingly one held by a majority of people in power. For them, the planet is not much more than a storehouse of resources for profits. For example, in March 2013, Representative Steve Stockman of Texas tweeted, in (mock?) celebration of Earth Day, "The best thing about the Earth is if you poke holes in it oil and gas come out." A similar-minded, well-known spokesperson stated, "God gave us the earth. We have dominion over the plants, the animals, the trees. God said, 'Earth is yours. Take it. Rape it. It's yours.'"[19] These are, perhaps, hyperbolic cases, but in fact we live in a world of such anthropocentric assumptions. We uncritically claim as our right the use of finite matters such as oil, coal, minerals, elephant tusks, whale blubber, and rhino horns. As Guy McPherson writes, "We see finite substances and the living planet as materials to be exploited for our comfort. We treat resources as our entitlement." It is a small step or maybe the same one—the result of what Horkheimer and Adorno would term the mythologizing of the Enlightenment—by which people exploit other people. Perhaps Freud addresses our current situation as well as anyone. Do our government and business leaders harbor a sort of worldwide death wish? Freud writes, "If we may assume as an experience admitting of no exception that everything living dies from causes within itself, and returns to the inorganic, we can only say 'The goal of all life is death,' and, casting back, 'The inanimate was there before the animate'" (*Beyond*). Yet psychoanalysis asserts "that at bottom no one believes in his own death, which amounts to saying: in the unconscious every one of us is convinced of his immortality" (*Reflections*).

Since the Renaissance, writers have engaged a sense of romantic mystery regarding collapsed civilizations. Perhaps the most widely read scholar on the subject today is Jared Diamond, whose 1989 article "The Present, Past and Future of Human-Caused Extinctions" identifies the unsustainable practice of human overhunting, introduction of nonnative species, and habitat destruction as the major elements in collapsed societies over time and around the world. The fauna of New Zealand, Madagascar, Hawaii and other Pacific islands, for example, "collapsed quickly after human arrival . . . . When one extrapolates from the studied islands to unstudied islands, one estimates that about one fifth of the species of birds that existed in the world a few thousand years ago have disappeared as a result of human activities on oceanic islands" (232).[20] Diamond argues against the idea that extinction caused by humans is "natural" and thus acceptable: "we make choices and alter the course of events around us" (235). Diamond would develop his ideas more fully in subsequent books and articles, as in his bestseller *Collapse: How Societies Choose to Fail or Succeed.* Diamond asks if unsustainable human practices—"ecocide"—will cause our current wealthy state to fail (3). He lists human-induced "environmental problems" as first among reasons for the collapsed societies he examines (487). Although Diamond does not address anthropocentrism directly, it is the basic, largely unexamined philosophical basis for such catastrophic choices.

Many literary works from antiquity to the present show an awareness of the possibility of human extinction and of a dwindled humanity. As I show in Chap. 2, Seneca is perhaps the most noteworthy ancient writer of ecocatastrophe, which is a major subfield of speculative fiction today, as well as a major framework in film and television. The scientific near-consensus that the world will someday exist with fewer or no humans is a logical basis bolstering the credibility of this *topos.* Humans are durable, clever, and adaptable, but there is a good chance that we are a temporary species. In his Templeton Prize speech, Sir Martin Rees said, "It won't be humans who witness the sun's demise: it will be entities as different from us as we are from a bug–either organic or silicon-based." As much as any recent fiction writer, the notion of human impermanence saturates the work of J.G. Ballard, whom I discuss in Chap. 8. Christopher Hitchens writes that Ballard is the only writer of which he is aware whose work is at home with the idea (353–54). Hitchens had apparently never read Cormac McCarthy or Margaret Atwood, though Ballard's speculative work is as counter to anthropocentrism as may be imagined, perhaps. But the serious questioning of anthropocentrism appeared in the West well before the twentieth century or Darwin, including among Christians and writers of other faiths (an idea that Hitchens would have surely contested). In fact, humans have nearly gone extinct before. In 2008, Stanford University researchers determined that humans almost became extinct 70,000 years ago ("Humans"). Stephen Jay Gould states that the geologic time scale is not appropriate for contemplating our own life and meaning: "we do not display cosmic vanity, but merely appropriate self-interest, when we choose to nurture and defend this particular little branch." Brief periods of mass extinction, he writes, are necessary for a vigorous tree of life: "an occasional catastrophic episode of mass extinction opens new evolutionary possibilities by freeing ecological space in a crowded world" (430).

Some, though relatively few, have called for direct action on alleviating the earth from its most destructive species. A movement calling itself the Voluntary Human Extinction Movement (VHEMT) espouses direct action to "the callous exploitation and wholesale destruction of Earth's ecology." The phasing out *Homo sapiens* "by voluntarily ceasing to breed will allow Earth's biosphere to return to good health. Crowded conditions and resource shortages will improve as we become less dense" (The Voluntary). A 1987 article by a right-wing deep ecologist appearing in the journal of the environmental organization Earth First!, "Population and AIDS," made waves by welcoming the disease as a valid means of reducing the earth's population.[21] Though not a transhumanist, John Gray, cited above, is among the more prominent early twenty-first century naysayers of humanity. The title of his book *Straw Dogs* (2002), a metaphor for humans, refers to an ancient Chinese ritual in which straw dogs were revered in a ceremony then trampled and thrown away (33–34). Humans are "plague animals" (12); wherever they go, Gray writes, "they carry the trash of litter of humanity" (150). Few writers attack our species and humanism so directly:

> *Homo rapiens* is only one of the very many species, and not obviously worth preserving. Later or sooner, it will become extinct. When it is gone the Earth will recover. Long after the last traces of the human animal have disappeared, many of the species it is bent on destroying will still be around, along with others that have yet to spring up. The Earth will forget mankind. The play of life will go on. (151)

Robinson Jeffers (whom Gray quotes) and Edward Abbey, among many others, precede Gray in charging our species with the grandest arrogance and error. Abbey writes that his first task as a young technical writer for Western Electric in New York was editing a manual entitled *How to Dispose of Human Sewage in Permafrost*, but he left the job almost immediately and drove straight to Arizona, quipping, "Never did learn how to dispose of human sewage (is there any other kind?) in permafrost" (*Beyond* 182).

Chapter 6 shows that Jeffers is perhaps the ultimate antihumanist poet, though he is certainly not alone in his vision.

Is antihumanism or, more central to works I discuss, the critique of anthropocentrism, a guise for mere misanthropy? A fair, all-encompassing answer to the question is impossible. Some writers suggest misanthropy, Swift, Mencken, and later Twain perhaps topping the list. But in other cases the misanthropic charge is a strawperson fallacy that disregards the scientific bases for posthumanist thinking. To equate the questioning of anthropocentrism with misanthropy is, on a basic level, dishonest, even if some writers seem to lend fuel to this judgment. As I see it, inquiry into our subject lies not in misanthropy but in an earnest search for meaning. Molière's great misanthrope, Alceste, finds his fellow men degraded and detestable, causing him to wish to "flee and find/Some desert land unfouled by humankind" (21). Yet many, perhaps most of the writers I discuss are represented not by Alceste but by Philinte, who responds to his friend's charge with, "Yes, man's a beastly creature; but must we then/Abandon the society of men?" It is this very human frailty, Philinte states, that leads us to philosophy (132). Antianthropocentrism and misanthropy are not synonymous, though there are certainly those who hold both views. Michael Zimmerman writes of the "passive nihilism" of some radical ecologists who "describe humans as the lowest possible animals, such as vermin, or even as cancer that should be eliminated" ("Nietzsche" 15). Another Molière character, Orgon, who has fallen under the cynical spell of the religious hypocrite Tartuffe, proclaims to his brother-in-law Cleante that the earth is a mere "dunghill of a world" (187). To equate the rejection of anthropocentrism with misanthropy is the result of either bad thinking or dishonesty. The recognition that humans are not the only species that matters has nothing to do with a hatred for fellow humans.

Ecological Apocalypse

At various points and especially in the last two chapters, this study will concern apocalyptic or postapocalyptic works, since these narratives so often revolve around humanity pared down greatly or almost extinct. Very few postapocalyptic works really concern a world without people because this would radically reduce the narrative possibilities, and few readers would want to read about a world absent humans. A kindred term is "ecocatastrophe," which *Merriam-Webster's* online defines as "a major destructive upset in the balance of nature especially when caused by the action of humans." This study is less about ecological apocalypse than literature that dramatizes a world without or with fewer humans—imagery, tropes, extended and brief, that explore the idea of a world without humans, or humans trimmed back significantly enough to be a much smaller factor in natural ecosystems. Since this elimination or trimming back does not seem to be achievable outside of a cataclysm of some kind, apocalypse becomes a major *topos* for an unavoidable ecocentric world.

Ecological apocalypse is a much-discussed idea in recent ecocriticism, and many works in the field (fittingly) quote Lawrence Buell, who states in his landmark *The Environmental Imagination* that "apocalypse is the single most powerful master metaphor that the contemporary environmental imagination has at its disposal" (285), and he demonstrates this idea with attention to Rachel Carson's *Silent Spring* and Leslie Marmon Silko's *Ceremony*. In his book *Ecocriticism*, Greg Garrard devotes a chapter to "Apocalypse," which he discusses as "an ecocritical trope" (94). Focusing on the trope as "misanthropic ecocentrism," he cites the character Birkin in D.H. Lawrence's *Women in Love*, who tells Ursula, "Man is a mistake, he must go" (qtd. in Garrard 101). Garrard states that the trope in the works of Nietzsche, Lawrence, and Jeffers "is ethically troublesome, as a truly ecocentric perspective would arguably be morally *neutral* regarding human impacts on the environment" (111), but he closes the chapter by stating, "Only if we imagine that the planet *has* a future, after all, are we likely to take responsibility for it" (116). For Garrard, environmental apocalyptic literature is "not about anticipating the end of the world, but about attempting to avert it by persuasive means" (107–08). Many of the writers I discuss have said as much about their own work functioning not as predictive or alarmist but rather as rhetorical and ethical. They seek to show us where we are right now and dramatize where they believe we are going if we stay on the current path. Their words are reminiscent of Jacob Marley's ghost presenting Ebenezer Scrooge with visions of where his world will end up if he doesn't change his avaricious ways. As we know, as a result of Marley's fortunetelling—which proved to not come true—Scrooge becomes a generous soul, vowing to live in the past, present, and future.

The term "ecological apocalypse" is redundant, since all apocalyptic events would be or would quickly become "ecological." Other than global climate change, which most scientists believe has been in-process for decades, nuclear war or an accidental nuclear discharge is our greatest catastrophic threat. Scientists researching catastrophes hold that we should have little fear that the sun will burn out (as dramatized in Lord Byron's poem "Darkness") for another five billion years or so. A super volcano, such as the one that erupted in Sumatra 73,000 years ago, could produce catastrophic results. Other sources of catastrophe—"the apocalypse"—include asteroids or comets, massive floods and tsunamis, earthquakes, massive storms, and disease or plague, though geologist Tony Hallam notes that it is virtually impossible for a disease to wipe out a whole species (7).

On smaller scales, whole societies have collapsed for a variety of reasons, including the Anasazi (Ancestral Puebloans) and the inhabitants of Easter Island, which was once a thriving civilization. Jared Diamond writes that the latter collapsed due to a self-inflicted ecological catastrophe. The inhabitants chopped down all of its trees and, setting off an ecological chain of disaster, lost its ability to sustain itself. As a consequence, the island has been uninhabited by humans for centuries (*Collapse* 79–119). On a larger scale, since around 1945, it has become clear that humanity is capable of fully rubbing itself out, though even one relatively minor nuclear event could wreak havoc on the planet, its climate and life. The Black Plague, the context for Boccaccio's *The Decameron*, gave late medieval Europeans a sense that the end of human time could be near. Other plagues ravaged Europe, including the "Great Plague" of 1665–1666, witnessed by Samuel Pepys and Daniel Defoe, which wiped out nearly a quarter of London, and the 1793 Philadelphia yellow fever epidemic provides the context for Charles Brockden Brown's novel *Arthur Mervyn*. Are we seeing the beginnings of an ecological apocalypse? There is evidence that this may be the case, though perhaps some have "cried wolf" too often in the past. Greg Garrard suggests that Paul Ehrlich follows but perhaps abuses the model of Thomas Malthus's 1798 *Essay on Principles of Population* in his bestselling book *The Population Bomb* (1969), whose dire predictions that the world would be pretty much over by 1980s did not materialize. Such hyperbolic rhetoric may finally be more dangerous than useful (Garrard 105).

Yet the Antarctic ice shelf is becoming increasingly unstable, Greenland is melting, and scientists believe that, for the first time in recorded history, the North Pole will soon melt. Other pressing environmental matters not unrelated to climate change can and perhaps will lead to crises of apocalyptic proportions—landfill shortages, the availability of drinking water, disease outbreaks, the dwindling bee population, and others. This book is not concerned explicitly with these very serious matters, but they underscore and bolster the logical bases of literary ecological apocalypse, as do a wide array of alarming situations observed by scientists. Tim Flannery writes that it is our beliefs about our relationship to each other and to the world, rather than our technology, that determines whether we show a Medean or a Gaian face (173).[22] We can either maintain a dog-eat-dog mentality or we can choose to understand our interconnectedness (22). If we proceed (as we largely have) to act as a "winner takes all species," the game is already over, and we will have become the destructive species par excellence (31). Flannery's ideas reflect why the accounting for the representation of anthropocentrism and its many challenges in literature matters. Our literature not only gives us a picture of changing views on our place in the world but provides alternative rhetorics to a hard anthropocentrism.

Despite the pain and horrors of real, imminent, potential, and imagined cataclysmic events, the apocalypse is a massive backdrop for popular entertainment. Alan Weisman's bestselling *The World Without Us* (2007) is a globetrotting rundown of dissolution drawing from anthropology, geography, as well as materials science. Based on an understanding of the Anthropocene, Weisman writes, "gradually, the asphalt jungle will give way to a real one" (28). He notes that humans have redesigned the atmosphere by CO_2, a phenomenon that otherwise requires erupting volcanoes and colliding continental plates to accomplish (39). Yet it is overpopulation that may be our undoing. One million new humans are born every four days; such numbers "will wax out of control until they crash, as has happened to every other species that got too big for this box" (271).

Less than a year after Weisman's book, a film then TV series *Life after People* (2008) appeared which speculates on nature's reclamation of the world from a single day to 10,000 years after people have disappeared through a nondescript means. A number of the show's CGI scenes resemble landscapes similar to those in Thomas Cole's *The Course of Empire* paintings (see Chap. 5). Within the first week or so after human disappearance, electricity fails, family pets starve, and much fresh food will go bad. After 10,000 years, virtually all traces of humankind will have vanished, even in great cities such as New York. Manmade structures such as the Empire State Building will have fallen, along with the Brooklyn and Golden Gate Bridges. By and large what is left has been around for millennia—the pyramids of Gaza and the Great Wall of China but also perhaps Hoover Dam and Mount Rushmore. The earth's literature will have decayed rather early on, and the earth's many radio and television signals, according to SETI calculations, will have dissipated within a light year or so, well before they arrive at the nearest star beyond the sun.

While the view that all human handiworks will not survive beyond ten millennia may be disconcerting for modern-day viewers–and is certainly a fatal blow to the conception of a universe centered around humanity—this is also an affirmation of life, which thrives well without humans. The Chernobyl site (only twenty years after evacuation) seems to confirm how quickly nature–flora and fauna–reclaim a place without humans, even though the area remains unfit for human habitation. The final narration in the film *Life after People* states, "If earth's 4 ½ billion years of existence were condensed to twenty-four hours, the passage of ten thousand years would be a fraction of a second. Man's time on the planet so far would be about half a minute long. So like an abandoned village on a global scale, the earth will move on without us. There was life before people; there will be life after people."

A number of popular feature films and television shows have broached the subject either in passing or as a central theme. *The Walking Dead*, based on graphic novels by Robert Kirkman, revolves around survival of pockets of humans, some benevolent and others not, after a plague has turned all but a relatively small number of humans into zombies. In Season 6 of *The Sopranos*, as Tony Soprano (James Gandolfini) lies in a hospital bed recovering from a gunshot wound, he reads aloud from a book on dinosaurs to his nephew Chris Moltisanti (Michael Imperioli): "If life on the planet is the Empire State Building, then humans have been present for the total of one postage stamp at the very top." Looking up, Tony asks, "You realize how insignificant that makes us?" Chris replies: "I don't feel that way," and the conversation ends ("The Fleshy Part"). In the first season of another HBO show, *True Detective*, Rust Cohle (Matthew McConaughey) holds a grudge against humankind resembling that of D. H. Lawrence's Birkin (though reportedly series creator Nic Pizzolatto's main influences are Thomas Ligotti and Emil Cioran). Rust explains his philosophical pessimism to his unsympathetic partner, Marty Hart (Woody Harrelson), as they ride together in their squad car. Human consciousness, he says, is a "tragic mistake in evolution." We're programmed to think we are somebody, but we're all nobodies. "I think the honorable thing for our species is to deny our programming, stop reproducing, walk hand and hand into extinction, one last midnight, brothers and sisters opting out of a raw deal" ("The Long Bright").

Steven Spielberg's 2001 film *A.I. Artificial Intelligence*, based on a story by Brian W. Aldiss, begins after floods caused by global warming have wiped out New York, Amsterdam, and other low-lying cities. The film's long closing sequence dramatizes a world without humans. A robot boy named David (Haley Joel Osment) and his (sex-robot) friend Gigolo Joe (Jude Law), both almost exact replicas of humans, set off for "the end of the world," Manhattan. As they enter the submerged city by stolen amphibicopter, only the forearm of the Statue of Liberty, with torch extended, remains visible, and beyond, lies the skyline, including the tops of the World Trade Center (the film was released less than two months before the 9-11 attacks).

Pixar-Disney's 2008 film *WALL-E* approaches a disturbing idea with charm. In the future, humans have trashed the planet; even the atmosphere (echoing images from the fiction of J.G. Ballard) is littered with satellites. The Earth no longer fit for life, the remaining humans have lived for seven centuries (presumably, beginning in the early twenty-first century) in a giant spaceship called Axiom, which is positioned far away from Earth, beyond the sun. Axiom is a mechanical world that the later H.G. Wells might have imagined, minus the explicit skepticism (a Disney film, its skepticism is embedded). In Axiom, everyone is obese, uncritically faddish, addicted to gadgets, but dysfunctional, kept alive and even governed by robots. Axiom is a mega-business and not the government per se, though apparently in this world they are one and the same. New York City is subject frequently to violent, Dust Bowl-like storms, and its skyscrapers are mixed with giant towers of trash, which the WALL-E (Waste Allocation Load Lifter Earth-class) unit was constructed to manage. The implications of the film are as radical as anything proposed in the opus of Edward Abbey: humans have cast pearls before swine, have given up their home for the sake of convenience, and as a result have exiled themselves for 800 years. Remarkably, the film grossed over $220 million in the theater (though its reported budget was $180 million). The film does, unsurprisingly, have something of a happy ending, but it is the twenty-ninth century, with much work to be done. Have humans learned their lesson? They have no memory of what exiled them from the Earth in the first place.

Peter Weir's *The Last Wave* (1977) is set within context of an Aboriginal Australian society established for around 50,000 years. A huge tidal wave at the very end of the film appears to be on its way to wiping out Sydney, populated largely by whites and founded by English colonists as a penal colony in the late eighteenth century. In Lars von Triers's *Melancholia*

(2011), characters await the life-ending collision of a rogue planet with the earth. The opening sequence includes beautiful but unsettling images of dissolution and falling objects, and it includes a shot of Bruegel's 1565 painting of a humanized winter landscape *Hunters in the Snow* blotted out by dark, falling rubble. The clinically depressed Justine (Kristen Dunst) is in a bad state after her marriage has fallen apart on her wedding night, but the promise of the apocalypse brightens her mood. As the clashing planet approaches earth, she tells her sister Claire (Charlotte Gainsbourg), "The earth is evil. We don't need to grieve for it. Nobody will miss it . . . we're alone . . . Life is only on earth, and not for long."

Hundreds of other shows and films are set after the apocalypse or a similar catastrophe. A partial list might include the following: the monster-science fiction film *Them!* (1954), in which the Edmund Gwynne character says, inverting Genesis, "the beast shall reign over the earth"; *On the Beach* (1959), based on the Nevil Shute novel set in Australia, where characters await fallout from a nuclear war; *La Jetée* (1962), a short film set in a ruined post-World War III Paris; *The Birds* (1963), Alfred Hitchcock's revision of the Daphne du Maurier story that holds implications for human dominance and the rebellion against it by nonhuman nature; *The Night of the Living Dead* (1968), George A. Romero's zombie film that includes several sequels; *The Planet of the Apes* (1968), in which humans have become a lesser species; *A Boy and His Dog* (1975), a post-nuclear war comedy-drama based on stories by Harlan Ellison; *Mad Max* (1979), directed by George Miller and set, along with its successful sequels, in a postapocalyptic Australia; *The Day After* (1983), set in rural Kansas after a nuclear war; *Threads* (1984), a documentary-style film set in Sheffield, England after a nuclear war; *Nausicaä of the Valley of the Wind* (1984), Hayao Miyazaki's animated tale of a postapocalyptic far-future world; *The Quiet Earth* (1985), filmed in New Zealand; *Waterworld* (1995), in which the polar ice caps have melted; *I Am Legend* (2007), based, along with two other previous film versions, on the Richard Matheson novel about the world decimated by a lethal virus; *The Road* (2009), based on the Cormac McCarthy novel; *The Hunger Games* (2012) and its sequels based on dystopian young adult novels by Suzanne Collins; and *Snowpiercer* (2013), a South Korean produced film in which the earth's remaining survivors live on a giant train after attempts to counter environmental crises backfire.

Unlike the films listed above, Michelangelo Antonioni's *L'Avventura* (1960) operates by realistic means to contrast images of the slow wheels of geologic time, represented by the Aeolian Islands, historical time (the old buildings and ruins of Italy and Sicily), and the ever-shifting follies of the shallow, well-to-do characters in this celebrated film. Antonioni establishes one of the film's chief recurring ideas of the old versus the new at the outset in the outskirts of Rome, with new, rather shoddy condominiums in the near background, some with construction scaffolding around them, while in the far background sits the dome of St. Peter's Basilica, completed in the early seventeenth century. The film's opening lines are spoken by a minor character, a builder, who says with regret, "Soon this poor villa will be smothered. To think there were woods here once." Motioning to the area, he says, "That'll all be houses. Then there will be nowhere to run." A well-dressed man, the father of a principal character, Anna (Lea Massari), replies, "Yes, there will be nowhere to run." Moments later a small rented yacht carries Anna, her boyfriend Sandro (Gabriele Ferzetti), Claudia (Monica Vitti), and a few other upper-class passengers past the jutting hulk of Basiluzzo Island in the Aeolian Sea, off the coast of Sicily. One of the female passengers observes about the scene, "Once the Aeolian Islands were many volcanoes," but her husband undermines her by reminding her that she said the same thing the last time they visited the area.

The passengers disembark on another uninhabited island, Lisca Bianca, upon which they observe millennia-old signs of human ruins, but they find no signs of Anna, who has disappeared mysteriously from the island. The cinematography further contrasts short and long time. An ancient city lies underneath the island's rocky surface, and one of the visitors drops an ancient vase which shatters on the rock. Much later in the film, searching for traces of Anna in Sicily, Sandro looks down from a church parking lot into a valley and remarks that a town lies below. Now Sandro's new lover, Claudia (the most introspective of principal characters) replies, "That's not a town; that's a cemetery." After Anna disappears, there is an initial flurry of search for her, but that dwindles, and she is almost forgotten in the latter part of the film, suggesting the human echo over the generations—lost, missed, mourned, then forgotten.

Narrative settings based on a world of fewer humans, whether by catastrophe or outright apocalypse, are central to other popular media outside of cinema and television. Many highly successful computer and console games—far too many to mention here—are also set in postapocalyptic settings, including *Fallout* and its sequels, *The Last of Us*, *Duke Nukem 3-D*, *Half-Life 2*, *Chrono Trigger*, *Pikmin*, *Ratchet & Clank*, and *Dying Light*. It could well be that our fascination with apocalypticism—in religion, in popular media, and in more highbrow forms—is the result of an in-born realization that we are a passing event on the planet. We seek narratives that try to make sense of, that contextualize, the human moment on earth, and the capacity for participating in these narratives range over a vast range of media, from the most popular to the most academic.

Yet this book is not about the apocalypses or "the" apocalypse, although the term is significant, sometimes central, in discussing many of the works that challenge and subvert anthropocentrism. The term apocalypse is ambiguous; it can mean "visionary" or, for critic Northrop Frye, "renewal and the imagery of heaven" (qtd. in Kinsley 47). Chapter 3 of Frank Kermode's *The Sense of an Ending: Studies in the Theory of Fiction* (New York: Oxford UP, 1966) discusses the history of the (mostly) ecclesiastical evolution of apocalypse. The word is derived from the Greek *apokalypsis*, which means "unveiling." The Vulgate Bible translated the word as "revelation," the nearest Latin equivalent. Most of the time the word denotes a "means of catastrophic destruction"—violence followed by peace. As the work of Morton D. Paley shows, romantic religious poets such as Blake work within an apocalyptic and millennialist framework influenced heavily by Milton, but so do unbelievers such as Shelley and Byron. The Bible and other religious texts often use literal and figurative nature imagery to dramatize apocalypse. Following biblical conceptions, many of the most important literary works, including *Piers Plowman*, *The Canterbury Tales*, *The Faerie Queene*, and *King Lear*, employ the idea in various ways. Yet the concept predates Christianity and Judaism. Norman Cohn writes that ancient myths of the Egyptians, Mesopotamians, and Vedic Indians share the idea "that cosmos has always been threatened by chaos and always will be, yet has always survived and always will" (*Cosmos* 65), but conceptions of apocalypse changed in the seventh century BCE with Zoroastrianism, for whom existence is "the gradual realization of a divine plan." Everything points toward a "glorious consummation" (77).

According to Cohn, Zoroastrianism later influenced the apocalyptic faith of Judaism (I Enoch and Jubilees in particular, but also Daniel, Isaiah, and others) as well as Christianity. Although wilderness for Zoroastrians was, as it would be for Israelites, an evil associated with chaos, humans have a heavy responsibility for warding off chaos, *draj* (disorder) (87). In the non-canonical Old Testament book I Enoch, following a Zoroastrian idea, animal souls will be called upon to accuse humans of wrongdoing (226). Often in the Old Testament, apocalyptic language regarding the fate of a sinful people employs images of wilderness and wild animals, such as in Zephaniah, which denounces the sinfulness of Judea and surrounding areas. The humble of the land may escape holy vengeance, but for others, the land, once full of human activity, will become "a desolation, and dry like a wilderness . . . a place for beasts to lie down in! every one that passeth by her shall hiss, and wag his head" (2: 3, 13, 15). For Zephaniah, the major sin of mankind is pride, which leads to rebellion against divine authority. Other apocalyptic sections of the Old Testament use similar language, including Daniel, Isaiah 2:9–19, 24:1–27, 13; Jeremiah 4:18–28; Ezekiel 38–39; in the Christian New Testament, the main passages are Matthew 24–25; Mark 13:14–20, 24–27 (the "little apocalypse"); Luke 21; I Corinthians 15: 42–55; 1 Thessalonians 4:13–5:11; 2 Thessalonians 2; and 2 Peter 3:1–13.

As Cohn shows in *The Pursuit of the Millennium*, in a variety of contexts over many centuries, the traditional belief in the second coming of Christ has served as the engine for social ambitions as well as hostilities. Many modern-day Christians believe the return of Christ as stated in the book of Revelation is imminent.[23] While Christians believe the end times are not only possible but likely, many of them tend to doubt that the end will arrive as the result of natural phenomena, including anthropogenic climate change. History is filled with doomsayers, most of them perhaps confusing their rather limited viewpoints with the largely unknown actions with what appears to be an indifferent universe. The case of the Millerites, who accepted the revelations of William Miller that Christ would return in 1843, is one of the most well-known examples. Perhaps the crying of wolf is an element in climate change denial, though it appears that skepticism toward it is based less on wolf-criers of the past than sheer ignorance. Yet even some respected environmental organizations tend toward alarmism, if not apocalypticism and dystopianism, in their appeals for donations. They appeal to the vanishing wilds ("Renew your Sierra Club membership for $25 to save the polar bears") and to the end of life as we know it. Are warnings by scientists about the catastrophic effects of climate change merely another form of Millerism? I don't think so.

Despite a long history of Miller-like prophets and Paul Ehrlich-like pronouncements, there is good reason for real alarm for action and change. In 2007, physicist and head of the NASA Goddard Institute for Space Studies James Hansen issued a sobering statement. Civilization was established during the Holocene, a period of stable climate that has existed for almost 12,000 years. The planet's warmness keeps ice sheets from developing in North America and Europe, but it is cool enough to maintain them on Greenland and Antarctica. Yet the warming in recent decades of 0.6 °C has driven up the temperature to the warmest level in the Holocene. Writes Hansen, "This warming has brought us to the precipice of a great 'tipping point.' If we go over the edge, it will be a transition to 'a different planet,' an environment far outside the range that has been experienced by humanity. There will be no return within the lifetime of any generation that can be imagined, and the trip will exterminate a large fraction of species on the planet" (qtd. in Speth 26–27).

Resistance to such alarms has been steady and loud. The Bush II administration, influenced heavily by (and comprised of) the religious right, would censor Hansen's work on the need to reduce the emission of greenhouse gases because it (supposedly) clashes with commerce.[24] Two years after Hansen's words, in a House of Representatives session on cap and trade (the control of pollution through economic incentives), Illinois Congress member John Shimkus denied the fact of climate change. Citing the Bible (Gen. 8:21–22 and Mat. 24), he stated that only God decides when the earth will end. Instances of public rejection of science by influential figures on the religious right are too numerous to chronicle.[25]

In contrast to many public members of the religious right, Jürgen Moltmann writes that the purpose of his work is

> not to go on distinguishing between God and the world, so as then to surrender the world, as godless, to its scientific "disenchantment" and its technical exploitation by human beings, but instead to discover God *in* all the beings he has created and to find his life-giving Spirit *in* the community of creation that they share. This view—which has also been called panentheistic (in contrast to pantheistic)—requires us to bring reverence for the life of every living thing into the adoration of God. And this means expanding the worship and service of God to include service for God's creation. (*God in Creation* xi-xii)

Ecologically-minded theologians such as Thomas Berry, Matthew Fox, Elizabeth Dodson Gray, Delores S. Williams, and Alister McGrath have, like Moltmann, emphasized creation or Sabbath-centered theologies.[26] Such statements are out of step with much mainstream Protestantism in the United States. Growing up in a conservative church, I was taught as a boy that although the earth is God's creation, it is not a holy thing in and for itself. Existence on earth is merely a temporal state. I was taught that, during our lifetimes, Christians would be raptured to a New Earth in Heaven while God's holy fire would destroy completely the old one. I was taught a clear dichotomy not only of body and soul but of earth and heaven. "Love not the world, neither the things that are in the world" was the banner used and which is still used to not only justify but mandate the exploitation, ownership, and profiteering of the earth's resources, Jesus being in many of my teachers' minds the CEO uber-capitalist of heaven. Any passing remarks about the beauties of the earth were tempered with suspicion and distaste for pagan admiration. I was taught that we are on the earth but not a part of it, and my early experience was not unlike that of millions of other people.

Notes

1. In *A Rhetoric of Motives*, Burke writes, "Rhetoric is rooted in an essential function of language itself, a function that is wholly realistic and continually born anew: the use of language as a symbolic means of inducing cooperation in beings that by nature respond to symbols" (43).
2. A March 2016 Gallup Poll shows that 65% of Americans believe in anthropogenic climate change and worry "a great deal" or "a fair amount" about it. See http://www.gallup.com/poll/190010/concern-global-warming-eight-year-high. Accessed 10 Dec. 2016.
3. The figure 40% refers to net primary production (NPP).
4. For more on weak anthropocentrism and its variants, see J. Baird Callicott, "On the Intrinsic Value of Nonhuman Species," Chapter 8 of *In Defense of the Land Ethic: Essays in Environmental Philosophy* (Albany: SUNY Press, 1989); Chapter 9 of Oelschlaeger; Bryan G. Norton. "Environmental Ethics and Weak Anthropocentrism." *Environmental Ethics* 8.2 (1984): 131–148; and Andrew Brennan and Yeuk-Sze Lo. "Environmental Ethics." *The Stanford Encyclopedia of Philosophy* Winter 2016 https://plato.stanford.edu.
5. Lynn White, Jr.'s often-cited article is "The Historical Roots of Our Ecological Crisis." *Science* 155 (1967): 1203–07.
6. See *The New Organon* (17). For more on Dominionism, see Michelle Goldberg, *Kingdom Coming: The Rise of Christian Nationalism* (New York: Norton 2006). On the anti-environmental activism of Dominionists, see Stephenie Hendricks, *Divine Destruction* (Brooklyn: Melville House 2005).
7. See also Nasr's *Religion and the Order of Nature* (Oxford: Oxford UP, 1996). See also Richard C. Foltz, et al., eds. *Islam and Ecology: A Bestowed Trust* (Center for the Study of World Religions, 2003). *The Encyclopedia of Religion and Nature*, edited by Bron Taylor, contains several brief sources

on Islam (among other world religions), including Kaveh L. Afrasiabi on "Islam and Post-Anthropocentrism."

8. Glacken includes Leibnitz along with Bacon and Descartes as chief agents in the changing views of man versus nature, which he argues is an outmoded concept. Though possessed of a "indefatigable, rich, and noble mind," Leibnitz "saw the earth as a divinely designed planet; he was untiring in his hopes for ameliorating the lot of mankind, and mastery over nature was beneficent, a mark of progress" ("Man" 130–31).
9. Kant writes elsewhere that "[o]ur duties towards animals are merely indirect duties towards humanity" and that rational beings alone have moral worth. Arguing from what is instrumentally wrong for people and not what is intrinsically good for nonhuman nature, Kant writes that cruelty to animals is wrong because it could desensitize one to inflict cruelty on people (*Lectures on Ethics.* Trans. Louis Infield. New York: Harper, 1963: 239–41).
10. Rejected by the De Lacey family, Frankenstein's monster responds with satanic vengeance on the natural environment: "I, like the arch fiend, bore a hell within me; and, finding myself unsympathized with, wished to tear up the trees, spread havoc and destruction around me, and then to have sat down and enjoyed the ruin." It is at this point that he decides to declare "everlasting war" on humans (111). Written by Ludovico Ariosto and based on a poem by Orlando Boiardo, *Orlando Furioso* (1516) is set during the war between Charlemagne and Agramante, the Saracen King of Africa. When Charlemagne's favored paladin Orlando discovers that a woman he loves, Angelica, has eloped with a Saracen knight, he rampages through Europe and Africa, destroying virtually everything along the way. See Ludovico Ariosto. *Orlando Furioso.* Trans. Guido Waldman. Oxford: Oxford UP, 2008.
11. Neil Badmington by implication also equates these terms in the opening sentence of his article for *The Routledge Companion to Literature and Science*: "Posthumanism marks a careful, ongoing, overdue rethinking of the dominant humanist (or anthropocentric) account of who 'we' are as human beings" (374).
12. See *The Civilization of the Renaissance in Italy* (1867), which has been translated into English in several editions. Burckhardt responded favorably to Nietzsche's writing but maintained distance from Nietzsche the person. See R.J. Hollingdale in *Nietzsche: The Man and His Philosophy* 2nd ed. Cambridge: Cambridge UP, 2001: 52.
13. Like Michael Zimmerman, Arne Naess has written about Heidegger's relationship with deep ecology, e.g., *Four Modern Philosophers: Carnap, Wittgenstein, Heidegger, Sartre* (1968). Rudolph Carnap, who fled Nazi Germany, was disturbed that his friend Naess would lump Heidegger in

with himself, Wittgenstein, and Sartre and write positively about him (*The Selected Works of Arne Naess Vol. 4.* Ed. Howard Glasser. Dordrecht: Springer, 2005: liii Note 16). Environmental philosopher Max Oelschlaeger is more skeptical: "Anyone who attempts to reconcile Heidegger's with Leopold's contributions to deep ecology finds the going rugged" (304).

14. Central to Foucault's critique is his concept of discursive practices in the human sciences: the "manifold relations, the open strategies, and the rational techniques that articulate the exercise of power" (*The History of Sexuality, Vol. 2: The Use of Pleasure.* New York: Vintage, 1990: 6). For Foucault, power and knowledge are two sides of the same process.
15. See Ligotti's collection *Songs of a Dead Dreamer and Grimscribe.* New York: Penguin, 2015.
16. The idea that *Homo sapiens* first appeared around 200,000 years ago is based on a *Nature* 17 Feb. 2005 article by Ian McDougall, Frank Brown, and John Fleagle. See http://www.nsf.gov/news/. Accessed 5 May 2015. A 2017 *Nature* article states that fossil evidence from Morocco places early humans at around 300,000 years ago (Jean-Jacques Hublin, et al. "New fossils from Jebel Irhoud, Morocco and the pan-African origins of Homo Sapiens" 8 June 2017).
17. The term "Anthropocene" was coined by biologist Eugene F. Stoermer in the 1980s and was popularized by Crutzen.
18. See NASA Earth Observatory. 22 Jan. 2016. http://earthobservatory.nasa.gov/IOTD/view.php?id=87359. For many in Congress, the solution to such findings by NASA scientists is to defund NASA's climate research arm.
19. Regarding Stockman, see http://www.theatlanticwire.com/politics/2013/03/steve-stockman-best-thing-about-earth/63399/. Accessed 4/14/13. The other quotation is from Ann Coulter. *If Democrats Had Any Brains, They'd Be Republicans.* New York: Crown, 2008: 104.
20. Diamond draws from Storrs L. Olson's article "Extinction on Islands: Man as a Catastrophe." *Conservation for the Twenty-first Century.* Ed. David Western and Mary C. Pearl. Oxford UP, 1989: 50–53.
21. See Christopher Manes, "Population and AIDS." *Earth First! Journal* 1 May 1987. One response to Manes is Murray Bookchin, "Social Ecology versus Deep Ecology: A Challenge for the Ecology Movement" in *Green Perspectives: Newsletter of the Green Program Project* (Summer 1987).
22. Flannery refers to Peter Ward's Medea hypothesis, which states that life is bloody and that species will exploit resources unwisely, leading to a collapse of the ecosystem (23).
23. A 2010 Pew Research poll found, "By the year 2050, 41% of Americans believe that Jesus Christ definitely (23%) or probably (18%) will have

returned to earth." A 2002 poll conducted by *Time* found that 59% of Americans believe the prophecies in the Book of Revelation will come true (Nancy Gibbs. "Apocalypse Now." *TIME* 160.1 (1 July 2002): 40).

24. See http://www.nytimes.com/2006/01/29/science/earth/29climate.html?_r=0. For more on Hansen's censoring by Bush II, see John Grant, *Corrupted Science: Fraud, Ideology and Politics in Science*. Wisley: Facts, Figures & Fun, 2007: 296–97.
25. Three more examples may suffice. Former Vice Presidential candidate and Alaska Governor Sarah Palin echoed Shimkus, denying the "political agenda" of scientists to "stymie development" (https://www.youtube.com/watch?v=aWR19LYSiTM). At a 2012 energy summit, former senator Rick Santorum rejected the "hoax of global warming." Humans, he said, were put on Earth "to use it wisely and steward it wisely, but for our benefit, not for the Earth's benefit" (http://www.huffingtonpost.com/2012/02/07/rick-santorum-global-warming-hoax_n_1260168.html). In 2012, Donald Trump, a staunch proponent of coal-burning energy (though not until his presidential campaign associated with the religious right), tweeted, "The concept of global warming was created by and for the Chinese in order to make U.S. manufacturing non-competitive" (https://twitter.com/realdonaldtrump/status/265895292191248385?lang=en). The fact that Trump included his name among other business leaders on a 2009 full-page ad in the *New York Times* calling for "a Clean Energy Economy" suggests a preference for political opportunism over science and global public health (see Philip Bump, "In 2009, Donald Trump endorsed action on climate change. Three months later, he disparaged it." *The Washington Post* 9 June 2016 https://www.washingtonpost.com/news/the-fix/wp/2016/06/09/in-2009-donald-trump-endorsed-action-on-climate-change-three-months-later-he-disparaged-it/).
26. See Thomas Berry, *The Dream of the Earth* (Berkeley: Counterpoint, 2015); Matthew Fox, *Original Blessing: A Primer in Creation Spirituality Presented in Four Paths, Twenty-Six Themes, and Two Questions* (New York: Bear, 1983); Elizabeth Dodson Gray, *Green Paradise Lost* (Warwick: Roundtable, 1979); Delores S. Williams, *Sisters in the Wilderness: The Challenge of Womanist God-Talk* (Maryknoll: Orbis, 1993); and Alister E. McGrath. *Re-Imagining Nature: The Promise of a Christian Natural Theology* (Malden: Wiley, 2016).

CHAPTER 2

The Earth as Pinprick: Some Early Western Challenges to Anthropocentrism

It is common to assume that the ancient Greeks and Romans were essentially anthropocentric in point of view. While this is partly true (as it is today), the ancients established important precedents that challenge and overturn this view, anticipating modern science and even Darwin and beyond. This chapter analyzes texts from the Presocratics to late antiquity to show how the questioning of anthropocentrism developed over roughly 800 years. Though written long ago, these works hold relevance for ecocriticism and for culture in general. Although few of the ancients developed the idea of biocentrism in depth, the seeds of the idea lie in their work, and it is obviously a necessary intellectual step toward objective, nonanthropocentric science and ethics.[1] This chapter, more than any other in this book, shows that the questioning or outright rejection of anthropocentrism is not a new idea, nor is it as extreme a position as some may imagine. The idea is rooted in the beginnings of the Western intellectual tradition. We have known some things about the folly of anthropocentrism for a very long time, well before Galileo and the early modern era. As Patrick Curry observes, "there is something ancient about an ecological ethic; it is more something we have forgotten than something we have never known" (12).

It is common to conclude that the writings in physics, ethics, and literature of the ancient Greeks and Romans, refocused and dogmatized by later Christians, established the anthropocentric mindset that justifies and dictates such practices today as the massive depletion of species, the burning of fossil fuels, and mountain-top removal coal mining. The ancients did tend toward anthropocentrism, just as consumerist-industrialist societies largely

B.L. Moore, *Ecological Literature and the Critique of Anthropocentrism*, DOI 10.1007/978-3-319-60738-2_2

do today, but there are also some important precedents among the ancients in interrogating and rejecting anthropocentrism. In fact, many ancient and medieval writers were open-eyed and, given their limited means for measuring and observing the universe, well-informed. Many of them did not believe that the earth is the most important part of the universe or that it is particularly unique. Some perceived the earth as tiny, almost irrelevant in the cosmos. Even the implications for humans in the astronomical writings of Claudius Ptolemy, whose geocentric system would be standard until Copernicus and Galileo, are still misunderstood frequently; "Ptolemaic" does not imply anthropocentric (Danielson 68).[2] Heliocentrism is central to the establishment of modern science, yet according to Archimedes, Aristarchus of Samos posited in the third century BCE that the earth rotates diurnally and revolves around the sun (see Gingerich 185–192).

Many distinguished writers have discussed the worldviews of ancients, including Clarence J. Glacken, Margaret Osler, and Max Oelschlaeger; my more modest aim here is a thumbnail view of a handful of chief figures to show that the anthropocentric mindset and—more importantly for my purposes—its interrogation, was established long ago, at least since the fifth century BCE. While the vast majority of ancient works are essentially anthropocentric, many of the ideas of the ancients crop up over and over into the present day, including the notion that the earth—and the humans that inhabit it—is a relatively tiny part of the universe, a "pinprick," as Seneca and other ancients described our cosmic position.

Ancient Greeks

Ancient Greek and Roman thinkers, beginning with the Presocratics, pioneered the seeking of rational explanations for the world not out of a Baconian desire to subdue or control nature, but to explore the place of humans in the greater world. The natural philosophy that began in sixth-century Miletus with Anaximander and Anaximenes represents the first attempt "to understand the phenomena of nature in purely physical or mechanical terms" (Kahn 2). The ancient Greek and Roman study of nature (*phusis*) made no claims about rigor; it did not, like modern science, exist for its own sake, but for what Pierre Hadot calls "a moral finality" (208). The detached distance experienced in the writings of some of the ancients provides a view of ourselves from above everyday life to show us the things that matter most—not luxury, power, fame, and the like, but philosophy. Hadot states that such a point of view is a sort of "exercise of death. One might say that this exercise has been, since Plato, the very essence of philosophy" (207). The tendency to strip ourselves of "the human" is constant through many ancient schools of philosophy (211).

Of course, the ancient Greeks and Romans did not use the words "anthropocentrism" or "ecology" in their writings. Yet many ancient philosophers anticipate the language of modern ecology and cast doubt on the centrality of humans in the world—sometimes within the same contexts. Plato and others up until the birth of modern science appropriated the careful observations of the Babylonians, who understood the movements of heavenly bodies as the purposeful activities of the gods. The work of the Greeks, from Thales to Plato's *Timaeus*, establishes not only Western philosophy but science and conceptions of nature itself. The cosmic scheme of Democritus and the atomists, writes Charles Kahn, "most fully anticipate the world view of modern science" (1, 2) and is an important precedent for Lucretius and his *De rerum natura* (*On the Nature of Things*), which was even more central in the birth of modern science.

Among the Presocratics, Heraclitus is most associated with the understanding that we live in a world of flux, while Parmenides is the philosopher most associated with the idea that we live in a world of permanence. (Both were born in the later sixth century BCE and died in the fifth century.) In Plato's dialogue *Cratylus*, Socrates says, "Heraclitus says somewhere that 'everything gives way and nothing stands fast,' and, likening the things that are to the flowing (*rhoē*) of a river, he says that 'you cannot step into the same river twice'" (402a). Based on this and fragments in other sources, later philosophers have concluded that Heraclitus believed that the world has an underlying unity, that this unity is dependent on a balance of opposites, and that change in one direction leads to change in another. Robin Waterfield comments on Heraclitus' skepticism toward humans relying too heavily on their senses: "there is nothing on the face of the world that we can securely grasp or base our moral opinions on; so we had better wake up and look to the underlying stability and unity of things" (34). This idea echoes in the work of many later writers, from Goethe, Wordsworth, and Emerson to A.R. Ammons, Gary Snyder, and Terry Tempest Williams. In a lost didactic poem, Parmenides lays out his thoughts on reality, which may be realized through reason alone, and not the senses. Speaking through a goddess, Parmenides states that "what is," is not subject to decay but is complete in itself, indivisible, and unchanging ("Parmenides" 397). W.K.C. Guthrie writes that Parmenides "was the exact reverse as Heraclitus. For Heraclitus, movement and change were the only realities; for Parmenides, movement was impossible, and the whole of reality consisted of a single, motionless and unchanging substance" (47). Guthrie's reading of Parmenides has been challenged as strictly monist, but those views persist in academia and elsewhere (see Palmer).

Among other Presocratics, Empedocles (c. 495-c. 435 BCE), called a "natural scientist" by later ancient writers, evokes protoecological unity in his *Physics* (extant only in fragments). Change is continuous, but, anticipating the Stoics, it is also orderly, personified as Love and Strife; and anticipating the Epicureans (he was a substantial influence on Lucretius), he suggests that change in nature is by chance (Barnes 136). Plutarch quotes Empedocles' statement that "there are effluences from all things that have come into being"; "not only animals and plants and earth and sea, but stones too, and bronze and iron, continuously give off numerous streams" (139). Irrational animals—his fragment cites hedgehogs—are better endowed than humans (150). It is unethical to kill living things, he writes, since there is "a law for all" (see Aristotle, *On Rhetoric* 1373b.2).

The Stoics would ridicule Empedocles, but he, again, anticipates the Stoics by writing, "There is a single spirit which pervades the whole world like a soul and which unites us with them" (Barnes 158). Even plants feel pain (159). Empedocles suggests a poetic, affirmative view of relationship between us and animals (161). In contrast to Aristotle's later understanding of final causes, laid out in Book II of *Physics* (especially 8.198b–199a), Empedocles suggests a natural selection in which only the most successful organisms would succeed. Though he does not provide evidence for such a theory, he does anticipate Darwin's *On the Origin of Species* by about two millennia.[3] Stoic physics, as it turns out, would be central in the questioning of anthropocentrism, even if Epicureanism espouses more centrally the rejection of this viewpoint.

Plato and Aristotle are often cited as the source of many misconceptions about the nature of the universe and the place of humans in it. Their views would largely persist until the observations of Galileo and the mechanical philosophies of Gassendi and Descartes in the seventeenth century. Plato and Aristotle held a teleological view of the cosmos, as did Anaxagoras before them, though their teleological bases are very different. Plato (and Socrates) turned away from the natural philosophy of many of the earliest Greek philosophers, though works such as *Timaeus* and *Laws* are attempts to explain the universe, mythically but also rationally—through *mythos* and *logos*. In *Phaedo*, we learn that "natural science" consisted (says Socrates) of a search for "the causes of everything; why it comes to be, why it perishes and why it exists" (96a). Related to his sense of political and ethical order, Plato rejects the materialism of the Presocratics in the thoroughly teleological scheme of *Timaeus*, which explains the universe as the ordered result of a single beneficent demiurge or "Craftsman" (29a–b). In contrast to the chance-governed materialism of the atomists, the demiurge imposed order on the cosmos, and the heavenly bodies are alive.

Aristotle rejects the teleological basis of his teacher Plato; for Aristotle, order has always existed. As he writes regarding "the four types of cause" in *Physics*, every natural process acts toward an "end or that for the sake of which a thing is done" (2.3). For Aristotle, the stationary earth is at the center of the finite universe, yet it is of no great size when compared to the fixed stars (*On the Heavens* 2.14). Following Plato, he writes that the divinity of the earth decreases from its circumference to its center—an idea one may observe in many later works, including Dante's *Inferno*, which places Satan at the very center of the earth's core. Plato's conception of the self-moving stars is impossible, though Aristotle has little to say about the "unmoved mover" in his treatise *On the Heavens*.[4]

The Stoics (beginning with Zeno of Citium in the fourth and third centuries BCE) re-enforced the anthropocentrism espoused by Plato and Aristotle: humans are in an elevated position. Christian thinkers would develop this idea in various ways, though they would largely abandon Stoicism for Platonism (by way of Plotinus) by late antiquity. Diogenes Laërtius, quoting the Stoic Posidonius (135–151 BCE), writes that the substance of the universe is "a complex of heaven and earth and the nature in them or a complex of god and humans and the things that come to be for their sake" (Inwood 52). Yet some Stoics diminish the importance of human affairs in their works and even veer closely towards a rejection of anthropocentrism. Animals are below humans for most Stoics, yet central to their philosophy is the idea of the unity of all things which is derived from the Presocratics. Not all users of the *topos* proceed uniformly, other than in diminishing man for whatever reason. Humans are small if one considers the great size of the world, though early Stoics tended to stress the notion that the earth is the absolute center of the cosmos. Stoic physics was an attempt to elevate the legacy of myth and legend "into science and philosophy, and to combine it with the cosmology of Heraclitus, seeing the world as flux and fire, conflagration and return" (Gillispie 182). The movement away from myth to science in the Hellenistic age goes hand in hand with a move away from teleological explanations of the universe and from anthropocentrism.

Ancient Romans

Cicero was a self-styled Academic, following Plato, but his understanding of the universe is clearly indebted to Stoic thought, especially by way of his friendship with Posidonius. In Book I of *The Laws*, Cicero lays out principles of natural law, including the ideas that the universe is ordered by rational providence and that man, a single species, stands between God and the animals; he is possessed of both animal needs and a godlike reason (1.1–57). As Niall Rudd writes, until recently "most people agreed with Cicero and the Stoics in assuming that man's dominion over the animals . . . was in accordance with natural law," oblivious to the idea that humans are capable of squandering resources, pollution, and anthropogenic extinction to such a high degree (Cicero, *The Republic* xxxi).

More the work of a Roman transmitter of Hellenistic thought than that of an original thinker. Cicero's *The Nature of the Gods* is chiefly a debate between Stoic, Epicurean, and Academic spokespersons. In the dialogue, he appears to side largely with a Stoicism that represents a strongly anthropocentric viewpoint. The dialogue wastes no time jumping into the main questions at hand: Do gods exist? If not, does the absence of gods create chaos? If there are gods, what is their nature? (In antiquity, natural philosophy and theology were inextricably tied.) Is the worship of the gods a "mere façade"? Cicero states that many esteemed philosophers hold that the universe is ordered, that all natural phenomena (weather, seasons, and the like) are created and "bestowed by the gods on the human race" (1.4).

Set forth first in the dialogue is the Epicurean case, presented by Gaius Velleius, who speaks, Cicero notes, "with the breezy confidence customary of Epicureans" (1.18). Stoics, Velleius charges, "prefer dreaming to reasoning" (1.19), and he ridicules the Stoic (and Platonic) notion that the whole cosmos is sentient and that the gods created the world for humans. He asks whether the gods made the world for all humans or only for the wise or for fools (1.23). Anticipating a similar argument in Hume's *Dialogues Concerning Natural Religion*,[5] he asks, rhetorically, why, if god made the world for humans, so much of the earth is uninhabitable for humans (1.24). He mocks the notion of fate, or *Heimarmene*, which holds that "every chance event is the outcome of an eternal verity and a chain of causation. How much respect can be accorded to this school of philosophy, which like a pack of ignorant old women regards all that happens as the course of fate?" By contrast, Epicurus (the founder of Epicureanism) brought freedom by teaching us to not fear superstition or the wrath of the gods (1.56)—an idea that would be developed by Cicero's younger contemporary Lucretius.

Responding directly to Velleius, Cicero has (Quintus Lucilius) Balbus deliver a rundown of the Stoic case, which is markedly anthropocentric. He quotes the Stoic Chrysippus in making a case for the existence of the gods, and in the process he shows the elevated state of man: if there are no gods, then nothing is better than man because he has reason (2.16). All things in the universe, Balbus continues, "have been created and prepared for us humans to enjoy," and "all that exists in the entire universe must be regarded as the possession of gods and men," including animals (2.155–56). Representing Cicero's own Academic skepticism, in Book 3, (Gaius) Cotta takes the edge off of Balbus' strict anthropocentrism by elucidating Chrysippus: if gods don't exist, nothing is naturally better than men, yet, said Chrysippus, to state that nothing is better than men is "supreme arrogance" (3.26). He mocks the Stoic claim that the gods have bestowed on man alone the faculty of reason (3.66). The gods, he continues, do not think of or care about human cities, much less humans themselves, which providence holds "in contempt" (3.93).

Weighing up the degrees of final causes and anthropocentrism, Cicero appears to back away from what one might call hard-line anthropocentrism. Like many ancients, including Seneca, discussed below, Cicero is genuinely interested in philosophy for its own sake and not as a rote set of laws that one should follow dogmatically without question. Other Cicero works underscore the impulse to move away from hard anthropocentrism, including *The Dream of Scipio*, which is the sixth and final book of *The Republic*. The importance to the history of ideas of *The Dream of Scipio* is difficult to overestimate; it would be a model for other writers, including Macrobius, whose *Commentary on the Dream of Scipio* was closely studied throughout the Middle Ages. The piece is derived from Plato's "The Myth of Er," which concludes his own *Republic*, and it is narrated by the second century (BCE) general Scipio Aemilianus, destroyer of Carthage, who recounts his service as a military tribune in Africa.

Set in a Platonic-Aristotelian geocentric universe of fixed stars and a motionless earth, Scipio falls into a deep sleep and is visited by the shade of his adopted grandfather, the Roman general Scipio Africanus, hero of the Second Punic War, who speaks of duty to the state and the younger

Scipio's future. In the dream Scipio is positioned in the stars, where he notes how small the earth is; the Roman Empire is only a tiny point on a tiny surface. The earth, says Africanus, is "in the middle of this celestial space" (6.15), and Scipio sees the "whole universe" which includes many stars that exceed greatly the earth in size. The earth, he says, "seemed so small that I felt ashamed of our empire, whose extent was no more than a dot on its surface" (6.16). Given this context of a motionless earth positioned in the lowest sphere, humans are incapable of hearing the Pythagorean music of spheres (6.18–19). The elder Scipio sees that his grandson is gazing on the tiny earth and says,

> "I notice you are still gazing at the home and habitations of men. If it seems small to you (as indeed it is) make sure to keep your mind on these higher regions and to think little of the human scene down there. For what fame can you achieve, what glory worth pursuing, that consists merely of people's talk? Look. The earth is inhabited in just a few confined areas. In between those inhabited places, which resemble blots, there are huge expanses of empty territory." (6.20)

He goes on to show how large areas of the earth are uninhabitable or occupied by non-Romans. "In the remaining areas of the east or west," he says, "who will ever hear your name?" (6.22). The fact that his speech avows a position of humility may at first seem anti-Roman, since the Romans, Cicero included, so revered their ancestors and statesmen. Posterity forgets us all eventually. On the brink of overthrowing Carthage, Scipio receives a giant dose of humble pie from his honored kin. This disavowal of earthly fame is purely Stoical by nature, and Scipio vows to live in the future with "a much keener awareness" (6.26).

Yet the elder Scipio also echoes the critique of Stoicism by the Epicurean Velleius in *The Nature of the Gods*: if the gods created the earth for human use, why is so little of it hospitable for human life? Far less a statement opposing anthropocentrism than an attempt to mortify human ambition, the idea that the earth is but a pinprick in the rest of the universe would be explored and expanded by later writers. Copernicus, and then Galileo, Newton, Herschel, Hubble, NASA space probes, and contemporary astrophysics would over time demonstrate the literal truth of this idea. Outside of Cicero, according to Polybius, after issuing the order to raze Carthage, Scipio Aemilianus says, "A glorious moment, Polybius; but I have a dread foreboding that some day the same doom will be pronounced on my own country" (38.5.21).

Lucretius is doubtless the key Epicurean figure, and he is also central in the rational questioning of anthropocentrism.[6] Lucretius writes in *De rerum natura* that the gods have no concern for humanity and the world, which is the result of the chance collision of atoms. Implicit in his view is a strong antianthropocentrism: "not for us and not by gods/Was this world made. There's too much wrong with it!" (2.181–82). Although Epicureans have an implicitly cyclical view of the world, Lucretius tends to ignore this. Near the end of Book 2, he envisions an exhausted earth, and like the writing of Seneca (see below), the book may be called ecocatastrophic, as in the famous lines in which he writes of a "shipwreck with spectator"—the enjoyment of catastrophe from a distance (2.1–2). In his celebrated book *The Swerve: How the World Became Modern*, Stephen Greenblatt shows the influence of Lucretius in the making of a modern, rational view of the cosmos. The reintroduction of Lucretius was central in the Renaissance and the birth of modern science. Writes Greenblatt regarding *De rerum* (4.1105+), "Human insignificance—the fact that it is not all about us and our fate—is, Lucretius insisted, the good news" (199).

In *Natural Questions*, Seneca (55 BCE–41 CE) is heavily Stoic in attempting to understand nature and its relation to ourselves. Harry M. Hine writes that it is surprising that Seneca wrote such a long work about physics, but he had, like Lucretius, an ethical aim: to lift the mind from narrow human concerns and survey the world as a whole—"the contemplation of the physical world complements moral action by shunning the full context of human action" (Seneca xvii). The focus throughout is on nature and natural events, but the human context is ever-present as he espouses the Stoic belief in the essential, inherent dignity and worth of all humans. Yet, veering toward materialism, the nonhuman is also possessed of inherent worth through the idea that even inanimate things have vital spirit, or *pneuma*; the earth itself breathes (6.16.1). We learn about ourselves by studying the rich variety of nature and is the means by which the mind can understand itself (7.25.1). We can learn the importance of things and interrelationships by looking outside ourselves to nature. So, he writes, shifting subjects, "let us inquire about terrestrial waters" (3.1.1). The Stoic notion that the earth is a living creature with a soul that can experience stress (6.14.2) holds some obvious comparisons with the twentieth-century Gaia theory of British engineer James Lovelock.[7]

As a result of his belief in the unity of all things, Seneca at times displays a protoecological and environmental awareness rather unusual for an ancient. He warns against the misuse of natural resources, and the pollution of the heavens is contrary to Stoic philosophy, which posits a tense relationship between the cosmos and its parts—the *pneuma* results in cosmic sympathy, something close to what we would term "ecology" today. He writes, "we cannot complain about god our maker if we have corrupted his good gifts" (5.18.13). The book holds many warnings against living for mere luxury and greediness; mines, for example, are used by the greedy (15.1). (One may temper such thoughts with the knowledge that Seneca, Nero's counselor, was extraordinarily rich. He has been attacked as a hypocrite over the centuries, but many have defended him.)

Seneca is, like Cicero, also open to other philosophies, including Epicureanism. *Natural Questions* rejects conventional Stoicism in its backing away from teleological explanations and anthropocentric reasoning. Seneca does not mention Lucretius or Epicureanism, but the book shares characteristics with Lucretius in presenting a rational explanation of events often attributed to malevolent or arbitrary gods—the attempt to replace fear with knowledge—and learn about ourselves along the way. The mind gains strength from the study and contemplation of nature and allows us to "stand above the abyss unflinching." Death is "not a great thing," being only a law of nature (6.32.5).

Again recalling Cicero in *Scipio*, Seneca, though involved centrally in Roman politics, appears to attack the imperialism of Rome. Earthly empires are insignificant when compared to the immensity of the cosmos. But he is less concerned with political power than with understanding the world philosophically, measuring "the world on its own scale," and showing "that the earth occupies just a pinprick" (4.11.4). For Seneca, almost everything in nature confirms the idea that "god did not make everything for human beings." Observing comets, for example, should show "How small a part of this vast creation is entrusted to us!" (7.30.3). He critiques the Etruscan teleological ascription of everything to a god; they say that clouds collide so that they will produce lightning. But Seneca is more given to the "coincidence" of natural events: the fates are not involved in the minutiae of nature (2.35.2). Near the end of the text, he wryly combines a Stoic view of death with the randomness of natural events: "But if you think that the turmoil of the heavens and the strife of the storms is being arranged for your sake, if the clouds are gathering and colliding and crashing on your account, if such powerful fires are being unleashed for your destruction, then count it a comfort that your death is so important" (2.59.12). He compares human operations with the activities of ants. Were they possessed of human intelligence, they would divide the world into provinces, yet this, again, only shows how, on the larger scale with which *Natural Questions* is concerned, kingdoms are only a "pinprick" (1.1.11).

Seneca's book is certainly one of the earliest works of ecocatastrophe written from a rational perspective, and this major theme seems to undermine Stoic anthropocentrism. Echoing Velleius in Cicero's *The Nature of the Gods* and anticipating (again) Hume, he asks why, if the gods have created the earth for our benefit, life is so marked by overwhelming events. Humans are "short-lived, frail creatures" who are subject to earthquakes (6.1.14). Sea torrents grow and wash the wreckage of nations into itself, containing human civilizations; afterwards, "remnants of the human race" cling to the heights (3.27.12). Yet (echoing Lucretius) earthquakes and the like don't happen because of a god: "these things," he writes, "have their own causes" (6.3.1). Natural catastrophes are very much in step with the unity of nature (3.27.1–3), a unity that extends to our own bodies, including human bleeding as a natural counterpart to the flow of earthly waters (3.15) and even farting and the emission of air from the earth (5.4.1–2).

The Renaissance translations of Pliny the Elder, a contemporary of Seneca, led to the study of natural history (Osler 132–33). Pliny's studies of a vast array of animals, geology, and other sciences (many of them premodern) were accompanied by his (more modern) observation of the human place in the greater world. In his massive *Natural History*, he casts doubt on the idea that Nature created everything for man. For Pliny, "It is ridiculous to think that a supreme being—whatever it is—cares about human affairs" (2.20). Like Roman writers before him, he attacks through ridicule the notion that humanity is at the center of the universe because we are so frail: "Pride of place will rightly be given to one for whose benefit Nature appears to have created everything else . . . man is the weakest among all living creatures" (7.4). Monkeys, he notes, are much like humans (11.246), and "only he who is always mindful of the frailty of man will weigh life in a fair balance" (7.44).

In the following century, Lucian (c. CE 115-after 180) was perhaps the most significant writer to subvert anthropocentrism. Born in Samosata (modern Turkey), he spoke Aramaic or Syrian but wrote in Attic Greek. A number of his philosophical dialogues attack human pettiness and the arrogance of the wealthy as well as what he saw as the foolishness of philosophers and, often, human nature itself, though he holds room for praise of those who live honestly, humanely. His satires are thus an important precedent for Voltaire, Swift, and Twain, and he was a model for Johannes Kepler in his *Somnium*, which defends Copernicus' theories on the movement of the earth.[8] While it would be a stretch to state that Lucian's dialogues reject anthropocentrism explicitly, they often veer in this direction. His *Charon or the Observers* was influenced by the Greek Cynics during the Second Sophistic, in which Romans looked back to Attic Greek writing of the Golden Age for inspiration. Cynics are (like Socrates) ascetics; virtue and self-sufficiency are the goals of life.[9] The Cynic Menippus, the third-century BCE Greek satirist, was an enormous influence on Lucian. The vanity of human aspiration and the mutability of fate are themes largely derived from Cynic thought (Lucian 13). Hermes states that if people were more aware of how fleeting human life is—as temporary as foam bubbles in a spring—they would live more reasonably and feel less grief over death. Cities die, says Hermes, as do even rivers (23). Elsewhere, Lucian, as is the wont of many satirists, takes a darker view of humanity. In *Dialogues of the Dead* a series of shades, including Diogenes, Menippus, Pythagoras, Socrates, and Alexander, along with gods associated with the underworld, Charon, Pluto, Cerberus, and Hermes, underscore human vanity.

His *Icaroneipuss or High above the Clouds* is a satirical dialogue, the title of which combines the names Menippus and Icarus. The former, the main speaker in the dialogue, flies on wings to the moon, then to Olympus, where he meets the gods, including Zeus. Largely a sideswipe at philosophers both well-known and obscure, Menippus pays a considerable amount of money for their learning but becomes confused with their talk of "first principles, final causes, atoms" (5). Determined to find wisdom, he flies to the heavens by attaching to himself the wing of an eagle and another from a vulture. On the moon, he notes that the earth is—echoing Cicero's *Scipio* —tiny, smaller that the moon. He is visited by Empedocles, who, burnt from the fire of Etna (into whose mouth he is stated in some sources to have cast himself), advises Menippus to flap only the wing of the eagle to acquire the bird's vision. In this way Menippus is able to see the minutiae of the earth and even individual humans. Greece, he observes, is very tiny, and the holdings of the greatest of landowners are merely the size of one of Epicurus' atoms. Visiting heaven, he hears the human prayers delivered to Zeus, but they are comically crude, mean, petty, and utterly selfish. Zeus pronounces philosophy useless and bids Hermes to send Menippus unceremoniously back to the earth.

Writers of the early Christian era, following Paul, generally stress the doctrine that since a Christian's true home is not the earth but the kingdom of heaven, "Our spiritual and worldly natures remain separate, and residence on earth is, in the end, inconsequential to the meaning of human life" (Peterson 34). Even suggesting that someone like Augustine is antianthropocentric demands serious qualification; within the spiritual context of *contemptus mundi* (contempt for the material world as suggested in e.g., I John 2:15), he downplays human significance in its illusory sense of earthly permanence in Book XI of *The Confessions.* He draws from both the book of Genesis and Plotinus in stating that the notion that a benevolent providence would only create a world which in itself is good goes against the grain of a harder, tragic belief system which perceives of a fallen, wicked earth as a place to plunder and despoil for materials and profit. Although his thinking is not compatible with modern science, Augustine, the most important of early church fathers, arrives at diminished role of humans in the world in ways comparable to that implied by modern science of the seventeenth century, the Enlightenment and Newtonian science of the eighteenth century and the theories of Darwin in the nineteenth century.

Commentary on the Dream of Scipio (*Somnium Scipionis*) of Macrobius, a Neoplatonist and probably a pagan, was tremendously influential in the Middle Ages through the Elizabethan period. He writes that humankind has, following Stoic doctrine, been frequently all but wiped out by a series of global catastrophes. In his cosmology, the matter that created the universe rises to the ether at the top of the cosmos, while earth, stationary and set at the bottom, is the repository for "the dregs and offscourings of the purified elements" (1.22.5). Macrobius writes: "Insignificant as [the earth] is in comparison with the sky—it is only a point in comparison, though a vast sphere to us" (2.5.10). He echoes many earlier works of literature, including *The Dream of Scipio*, by stating that only a fraction of the earth is temperate and habitable for humans.

At the beginning of the Middle Ages, Boethius' *The Consolation of Philosophy*, written in prison in 524—he would be executed shortly after for alleged treason against the king—became the cornerstone of medieval humanism. Bridging classical traditions with Christianity, the work is about finding happiness and meaning amidst a world of human suffering. Man has a "need to explore and reveal Nature's secret causes"—the work makes use of many nature images and metaphors (see Poem 6)—but now the speaker's mind is deadened, and he can only stare at the "dull earth" (1

Poem 2). In a seeming understatement, man is "no small part of [God's] great work," but he is subject to Fortune (1 Poem 5). The anthropocentrism in the work unfolds slowly but is never absolute; the fruits of earth were "given to animals and men," and God wished humans to be above "all earthly things" (2 Prose 5). It is only the human race that stands erect and (evoking *contemptus mundi*) looks to heaven "despising the earth" (5 Poem 5). Yet when humans forget who they are, they become like beasts (2 Prose 5 and 4 Prose 3). Echoing Cicero's Scipio, human ambition is an empty thing: "the whole circumference of the earth is no more than a pinpoint when contrasted to the space of the heavens." The earth has comparatively "no size at all"; the habitable lands are an "insignificant area on a tiny earth" (2 Prose 7). Through the twelfth century, Boethius, along with Plato's *Timaeus* (by way of Calcidius), Seneca, and Macrobius, would dominate scientific thought (Dales 37).

Granted, none of the ancient writers I have discussed hold what one might call a biocentric or ecocentric viewpoint. Although many thinkers intuitively as well as rationally anticipate such a view, its grounding in science would require evidence beyond the reach of Aristotle, whose physics dominated the West from the thirteenth century until the Renaissance. The astronomical observations of Galileo and the philosophy of Descartes overturned Aristotle. In the nineteenth century, the watershed work of Charles Darwin, and finally the rise of ecological science and environmental ethics in the twentieth century would make biocentrism possible. The tragic effects of the Anthropocene would underscore the importance of challenging anthropocentrism and work their way, eventually, into legislation such as the United States Wilderness Act of 1964 and various worldwide attempts (with failures and successes) to control the burning of fossil fuels.

The rise of Christendom would present another story beyond the scope of this chapter, but, like the ancients, one labels the early Christian thinkers purely anthropocentric at much peril. Doubtless, the Bible has been throughout history used much more often to justify the exploitation of the earth than its good stewardship, as espoused by such writers as John Muir, Wendell Berry, Annie Dillard, and Terry Tempest Williams. However, portions of it problematize anthropocentrism. For example, Job replies briefly, meekly (in Hebrew Wisdom parallelism) to Yahweh's heavily poetic speech rife with rhetorical questions,

I am worthless. What can I say back to You?
My hand I put over my mouth. (40:3)

And in the New Testament, Paul writes that "the earth is the Lord's, and the fullness thereof" (I Cor. 10:26, KJV). Many early church fathers, including Arnobius and John Scotus Eriugena, as well as the most important Jewish theologian of the Middle Ages, Moses Maimonides (though an ardent antagonist of Epicureanism), respond to anthropocentrism negatively.[10] Whatever the case, this chapter has provided sufficient evidence to challenge the notion—all but a commonplace for some—that the ancients were absolutely anthropocentric in outlook. Far from it.

Notes

1. On the relationship between stoic science and ethics, Lawrence C. Becker writes, "When we say ethics is subordinate to science we mean, among other things, that changes in our empirical knowledge are likely to generate changes in ethics. When the best science postulates a cosmic telos, as it sometimes did in antiquity, so does stoic ethics. When the best science rejects the view that the universe operates teleologically, in terms of something like human purposes, and suspends judgment about whether cosmic processes have a de facto end, convergence point, or destination, so does stoic ethics" (*A New Stoicism*. Princeton: Princeton UP, 1997: 11).
2. Michael J. Crowe notes the complexity of Ptolemy's theories in *Amalgest* and other works: he "had not one, but rather a number of systems—one for each of the main bodies of our system" (*Theories of the World from Antiquity to the Copernican Revolution*. 2nd ed. New York: Dover, 2001: 43).
3. In late editions of *On the Origin of Species*, Darwin cites both Empedocles and Aristotle on evolution in *Physics*, though he doesn't acknowledge that Aristotle's views are in contrast to Empedocles'.
4. In *The Inferno*, Dante places Aristotle in his Limbo, along with Socrates and Plato, Democritus ("who strove to show/That the world is chance" (4.120–21)), Diogenes, Seneca, Averroes, etc. Satan is frozen in Lake Cocytus at the center of the earth in Canto 34.
5. Hume undermines the argument by design (in part) by arguing that the earth contains too many flaws to have been created by an omniscient, beneficent deity.
6. Virgil was also an Epicurean, and he is certainly the great Roman writer, but his poetry is less directly and didactically concerned than Lucretius' with Epicureanism.

7. Lovelock's Gaia theory, named after the ancient Greek representation of the Earth, argues that all living entities, from simple (a virus) to complex (a whale), form a single living entity. See Lovelock's *Gaia: A New Look at Life on Earth* (New York: Oxford UP, 1987).
8. Another important source for Kepler is Plutarch's *The Face on the Moon*, which the astronomer read in Greek. See John Lear's Introduction to *Kepler's Dream*. Berkeley: U of California P, 1965. 84.
9. The word "cynic" is from the Greek word for dog: Diogenes the Cynic was, by tradition, "as shameless as a dog." See *The Cynic Philosophers: From Diogenes to Julian*. Trans. Robert Dobbin. London: Penguin, 2012. xi–xii.
10. The second chapter of Peterson's *Being Human* discusses an orthodox Christian position on "human exceptionalism." See also Glacken's *Traces on the Rhodian Shore* and B. Moore, 55–62.

CHAPTER 3

Lowering the Human Throne: European Literature to 1900

This chapter traces the movement in European literature, but especially British poetry, from the Middle Ages up to the twentieth century that dislocates humans from atop the *telos* to regions somewhat lower. I do not claim that this realization had much effect in moderating how humans treat the earth; any dislocation is poetic, artistic, largely undergirded by science, and with a rhetorical effect that is difficult to trace. Nor do I make claims to completeness in my survey; I can only hope to discuss a relatively few highlights to form a picture of the evolution of a collective consciousness directed largely by scientific inquiry. Few of these works are purely what one might call antianthropocentric, yet the idea emerges at times, sometimes indirectly. Many of them evoke the temporality of earthly human life as a means of contrasting and instructing readers in the hope of eternal heaven, while others, following ancient models, seek to mortify ambition or express the age-old idea of the brevity of human life. Almost all of them, though, diminish the centrality of humans in the universe in some way.

Many of the works in this chapter intersect my theme of antianthropocentric literature with material ruins, especially Roman, and they underscore not only the brevity of life but also the (in)significance of a being whose presence is, individually but also collectively, and especially in geological and universal-spatial terms, so temporal. Writes Robert Ginsberg, "The ruin comes home to the primacy of Mother/Father Earth" (56). The effects of exterior, material ruins on the human mind are myriad, but it is difficult to look at a ruin without seeing it in its relation to human time; as Walter Benjamin writes, "in the ruin history has physically merged

B.L. Moore, *Ecological Literature and the Critique of Anthropocentrism*, DOI 10.1007/978-3-319-60738-2_3

into the setting" (*The Origin* 177–78). Georg Simmel also saw ruins in terms of the human in contrast to nonhuman nature. In ruins, "natural forces begin to become master over the work of man: the balance between nature and spirit, which the building manifested, shifts in favor of nature. This shift becomes a cosmic tragedy which, so we feel, makes every ruin an object infused with our nostalgia" (259).

The ruin is a main trope for evoking human transience, the precariousness of civilizations, and a humble sense of one's relevance before human constructions that have been in existence for hundreds and in cases thousands of years, within the context of the average human lifespan of seventy-five years. It is also difficult to discuss ruins without, to some degree, sidestepping the discussion of what Michel Makarius calls ruins stereotypes: "the lyricism of desolation and empty wastes, the meditation on time past, the collapse of empire, the brevity of human life," assuming the form of "irresistible decay" (165). The understanding of temporality seems to be a necessary precondition for questioning and rejecting anthropocentrism.

Robert Ginsberg writes of a "romantic theory of ruins" that states, "we too are subject to ruin. The reminder that all things pass away renders our present sober/somber/solemn" (315). François-René de Chateaubriand, Henry James, and Sigmund Freud wrote of the psychological connections between the human mind and ruins. Writes Ginsberg, "For Chateaubriand, our mortality is in the balance. He will have Catholicism rescue us from it. For James, aesthetic pleasure wins out, even when we encounter suffering. . . . For Freud, hidden mental entanglements drag us down and muck up our life. He will liberate us by a therapeutic humanism that brings to light our mind's ancient ruins" (362). James's statement in *Italian Hours* ("Roman Rides") concerning the "perversity" of ruins has been taken out of context by some commentators. He does not claim that the observation and even enjoyment of ruins is perverse but rather the conception of ruins as "sentient." James's critique is thus akin to Ruskin's pathetic fallacy. James initially imagines that some of the villas "have an indefinitely sinister look," but he steps back to comment on this line of thought:

> To delight in the aspects of *sentient* ruin might appear a heartless pastime, and the pleasure, I confess, shows the note of perversity. The somber and the hard are as common an influence from southern things as the soft and the bright, I think; sadness rarely fails to assault a northern observer when he misses what he takes for comfort. Beauty is no compensation for the loss, only making it more poignant. (441)

Yet such "perverse" thoughts about ruins instill much writing on the subject, old and new. James would employ the ruins of Rome in his novella *Daisy Miller*, the title character of which rejects propriety, remains in the Colosseum until midnight with her suitor Giovanelli, and consequently expires from Roman fever. Hence Daisy proves to be one of many other short-lived spring flowers which decorate the ruins for a short span then goes away, leaving the ruins to continue their slow decay.

The Middle Ages Through the Neoclassical Era

With few exceptions, ruins poetry would not appear until the Renaissance, yet the brevity of human life against the context of a lasting natural world and, more particularly, the eternal heaven, is an idea in the earliest of English poetry. One of the recurring themes in the Old English *The Exeter Book* (copied ca. 940) is the temporality of earthly life, where moth and dust corrupt, as compared with the city of God. Several of the poems work didactically, preaching *contemptus mundi*. *The Wanderer* is an elegy and one of the book's most celebrated poems. A meditation on fallen warriors, the sage/speaker of the poem concludes that on earth, "wealth is ephemeral; here a friend is ephemeral; here man is ephemeral; here kinsman is ephemeral; all this foundation of earth will become desolate," and so he must seek "grace, consolation from the Father in heaven, where for us all the immutable abides" (Bradley 325). "The Ruin," the earliest English poem about ruins, is itself, perhaps fittingly, a fragment and partially unintelligible. The text of the extant manuscript, in its meticulous description of a once great but now fallen city, is likely intended as a picture of earthly mutability as a contrast with the immutable City of God: "There are tumbled roofs, towers in ruins, high towers rime-frosted, rime on the limy mortar, storm-shielding tiling scarred, scored and collapsed, undermined by age. An earthy grasp holds the lordly builders, decayed and gone, the cruel grip of the ground, while a hundred generations of humanity have passed away" (402).

In what is among the earliest of all English pilgrimage poems, *The Seafarer*, lighting rarely on land, the title character prefers the song of the swan, the cry of the gannet, the call of the curlew to human laughter, "the sea-mew's singing in place of mead-drinking," and the cuckoo who "serves warning by its mournful cry." Echoing ancient eclogues (and anticipating latter-day American country music), the speaker doesn't expect city people content with society and worldliness to understand him, yet he evokes the beauty of cities along with that of the woodlands and meadows: "Great is the awesomeness of the ordaining Lord, for this world will pass away" (334). Repeatedly, the poet praises humility and moderation, yet the things in nature are not ends in themselves but rather temporal things that are, at best, signs of God's creation.

In the later Middle Ages, several important poets echo Cicero's *Scipio* in representing the smallness of the earth within a heavenly—and universal—context. Dante, in *Paradiso* 22.133–55, guided by Beatrice, ascends into the heaven of the fixed stars, looks at the seven spheres below him, and smiles at the insignificance of the "little patch of earth" (22.151). In Boccaccio's *Theseid of the Nuptials of Emilia*, following his death, the ghost of Arcites flies toward eighth heaven and admires the order of the planets and the sweet heavenly sounds, and then, gazing downward toward "the small terrestrial sphere around which rotated the sea, the air and above them the fire and everything deprived of value compared to the heavens . . . it laughed in itself of the sorrowful laments of the Lernean people, and decried deeply the vanity of humankind that, madly obfuscated in their minds by an obscure blindness, pursue the false appeal of the world" (11.1–3). A similar idea appears in some of Chaucer's poems. *The Parliament of Fowls*, a Valentine's Day poem, begins with the lines, "The lyfe so short, the craft so long to lerne" (1), and the poem's first several stanzas are a commentary on Cicero's *The Dream of Scipio*. Scipio shows Scipio the Younger the Milky Way and "the lytel erthe that here is,/At regard of the hevens quantite" (57–58). The earth, he continues, is "lyte,/And dissevable and ful of harde grace" (64–65). In Chaucer's *Troilus and Criseyde*, after Troilus is killed by Achilles, he ascends to the eighth heaven and looks at "This litel spot of erthe" (1815) and rejects human vanity.

C.S. Lewis argues that, for writers of the Middle Ages and Renaissance, the smallness of our species was a given. Alanus ab Insulis (Alain de Lille), for example, compares the universe to a giant city. We on earth are on the outer edges, the suburbs. Alain reflects in part Calcidius, the influential fourth century commentator on Plato's *Timaeus*, who states that humans watch "the spectacle of the celestial dance" from the fringe: "The Medieval Model is, if we may use the word, anthropoperipheral. We are creatures of the Margin" (Lewis, *The Discarded* 58).[1] Dante, Lewis writes, shows more clearly than any that "the spatial order is the opposite of the spiritual, and the material cosmos mirrors, hence reverses, the reality, so that what is truly the rim seems to us the hub" (58). Of course, this medieval prescientific context is very different from the anthropocentrism rejected by more contemporary writers and philosophers: the cause for the reduced significance of humankind is very different, but the effect is similar. E.M.W. Tillyard writes, "Far from being dignified and tending to an insolent anthropocentricity, the earth in the Ptolemaic system was the cesspool of the universe, the repository of its grossest dregs" (39). And discussing pre-Copernican conceptions of the cosmos, Steven Shapin writes, "Although human beings, and their earthly environment, were understood to be the unique creations of the Judeo-Christian God, compared with the heavens and a heavenly afterlife the earth and earthly existence were regarded as miserable and corrupt, and the actual center of the cosmos was hell" (24).

Space allows only a brief sketch of the rise of literary-artistic interest in ruins. Some degree of fascination with ruins traces back far into history. Second-century BCE Romans, for example, contemplated the ruins at Carthage and then proceeded to build their own city on top of the old one. Part of the credit for developing the Renaissance taste for ruins lies with Petrarch. As the greatest early modern translator of Livy and Cicero, Petrarch initiated interest in Rome as "an entity in itself, a worthy object of study" (Makarius 58). Regarding his first visit to Rome in 1337, he wrote in a letter of being "overwhelmed . . . by the wonder of so many things and by the greatness of my astonishment." He had wondered if his high expectations for Rome established from books would live up to seeing it in person, but he says that his visit "diminished nothing and instead increased everything. In truth Rome was greater, and greater are its ruins than I imagined. I no longer wonder that the whole world was conquered by this city but that I was conquered so late" (113). In a later letter to his friend Giovanni Colonna, he reminisces about their walks through the "broken city" where "the remnants of the ruins lay before our eyes" (294), though he does not include any specifics.

Ruins have long existed as *memento mori*—reminders of the brevity of life and the vanity of human ambition. According to Michel Makarius, the idea of ruins emerged into human consciousness in the fifteenth century (17). Often cited as the first literary work that celebrates ruins, as opposed to those invoking *contemptus mundi*, the *Hypnerotomachia Poliphili: The Strife of Love in a Dream* (1499), attributed to the Franciscan monk Francis Colonna, shows enthusiasm for the Latin culture of the Quattrocento. In the work, Poliphili pursues his beloved Polia, but is sidetracked by ruins, which he finds chaotic but filled with meaning; his cries of "wondrous terror" and "delightful horror" while in the ruins suggest what would become the Sublime in later centuries (Makarius 15). Writes Makarius, "The historical significance of ruins is necessarily based on the conception humankind entertains of temporality," a concept that was impossible, he says, before the Renaissance artists of the Quattrocento (8). Ruins are a staple for many Renaissance painters, including Sandro Botticelli and Albrecht Dürer, both of whom represent the Adoration of the Magi in ruins settings. Beginning in the sixteenth century, the English Reformation would bring about the dissolution of monasteries, the whitewashing of church wall paintings, and the like, but it would also lead to an aesthetics of the ruin (Groom 28).

A century or so later, early British neoclassical writers such as Thomas Burnet embraced ruins enthusiastically (albeit natural, postdiluvian ones) as a theme and gave rise to an appreciation for irregularity in nature and the Sublime.[2] Later in the eighteenth century, French writers and painters extended the grammar of ruins aesthetics. Writing about the Salon of 1767, Diderot expresses admiration for the ruins paintings of Hubert Robert, but also disappointment in the painter's ignorance of a "poetics of ruins." Ruins evoke in Diderot sublime ideas:

> The ideas ruins evoke in me are grand. Everything comes to nothing, everything perishes, everything passes, only the world remains, only time endures. How old is this world! I walk between two eternities. Wherever I cast my glance, the objects surrounding me announce death and compel my resignation to what awaits me. What is my ephemeral existence in comparison with that of a rock being worn down, of a valley being formed, of a forest that's dying, of these deteriorating masses suspended above my head? (*Diderot* 198–99)

Yet the contemplation of ruins, he states, also provides a kind of comfort, make him feel "freer, more alone, more myself, closer to myself" and cause him to "probe my own heart" (199). Painting and writing about ruins would develop largely along the lines established in Diderot's Salon commentaries, a foretaste of romantic thinking. A few tumultuous decades later, Chateaubriand, discussed below, saw the relationship between ruins and humans as an analogy regarding our own brief lives.

An uptick on ruins as a poetic theme occurred in the sixteenth century from writers in a variety of European countries. Baldassare Castiglione was among many humanists who were interested in archeology. His sonnet "Superbi colli, e voi sacre ruin," written around 1503, the year of his first visit to Rome, apostrophizes the city, its "Monuments to glorious and joyful triumphs,/All reduced to a little ash" that now "tell a mean tale to the common herd." The poem concludes with the secular comfort that perhaps time will bring an end to the speaker's own sufferings. Even more important, Joachim du Bellay's (d. 1560) *The Antiquities of Rome* (*Les Antiquities de Rome*) inaugurates a poetry of ruins. This sonnet sequence, formal and inspired heavily by Petrarch, focuses little on the ruins themselves and more on an imagined future when France will succeed the hegemonic rule of the pope to create a new Rome. The sonnets also glorify ancient Rome and invoke the relative immortality of great art, including du Bellay's own verse, as in the beautiful concluding poem (No. 32). Poem 23 warns against hubris: "in a nation with too much leisure, ambition is easily born, like illness in an unhealthy body" (10–11). But du Bellay feels no need to "repeat the commonplace that everything beneath the moon is corruptible and subject to death" (9.10–11). At its beginning, poetry about ruins explicitly evades its most common expression of the brevity of human life! The addendum to *The Antiquities*, *A Dream of Vision*, consists of fifteen sonnets with heavy symbolism resembling the book of Revelation with strains of Ecclesiastes: "See how all is nothing but vanity" (1.1.8). Many of these sonnets are structured similarly: the speaker observes the vision of a giant precious artifice, usually representing Rome in some way, but by the end of the poem, the artifice crashes down, as in a barbarian invasion.

The greatest of all early modern English poets to write about ruins, Edmund Spenser was inspired heavily by du Bellay in *The Ruines of Time*, which opens *Complaints* (1591), and the sonnet series *Ruines of Rome: by Bellay*. The former work is filled with the poet's sense of loss and change, but it is tempered with the assurance of heavenly immutability and the immortalizing power of poetry. Evoking Ecclesiastes more than the New Testament, the poem bemoans the "vaine worlds glorie, and unstedfast state" (43). Everything that lives knows little happiness; living things "crying creep out of their mothers woomb,/So wailing backe go to their wofull toomb" (48–49). Spenser suggests that the Thames once flowed through Rome but was diverted and proceeded to transport the Roman ruins to England. The poem reaches a conclusion similar to that of *The Wanderer*: the world is filled with grief and "sad spectacles" (576). All the ruins the poet has seen underscores the conclusion that is spoken by a heavenly voice near the end of the poem:

> . . . by ensample see,
> That all is vanitie and griefe of minde,
> Ne other comfort in this world can be,
> But hope of heaven, and heart to God inclinde;
> For all the rest must needs be left behind . . . (582–86)

Spenser follows the traditional Christian model of a heavenly solution to the problem established by Boethius, and this model would predominate through the Elizabethan era and beyond. Anything outside the construct of the Great Chain of Being, which was accepted into the eighteenth century, is no matter for concern, or, as Tillyard writes, "The apparently superfluous could be put low down on the ladder of creation, and the matter was settled" (31). Remove the heavenly solution and one has a rather modern poem. In this sense, ancient writers such as Seneca and Lucretius are more "modern" than the chief poets of the English Renaissance.

Spenser probably intended the *Two Cantos of Mutabilitie* as the core of an additional book (the seventh) of *The Faerie Queene*. Although unfinished, this remarkable poem reflects the colonialist context of late sixteenth-century Ireland, which was, for Spenser and the English colonists, falling increasingly into chaos. (In 1598 Spenser's castle would be burned by rebelling Irish in the Nine Years' War, forcing him to move to London, where he died months later.) Yet if the poem were merely about colonial anxieties and fears, it might hold little more than historical interest. The poem also reflects universal, existential questions about permanence and change that most if not all humans experience, whether it be within the ignoble context of suppressing rebelling natives or of life itself. Is change—or chaos, ruin, decay, instability—the only sure thing in life? Spenser examines this seemingly cold fact in the person of Mutabilitie, who claims sovereignty over the universe, on center stage to address the question.

Mutabilitie delivers her complaint at a gathering of natural and supernatural beings unseen since the marriage of Thetis and Peleus on Arlo Hill, near Spenser's Irish estate. The judge of the hearing—following the model of Chaucer's *A Parliament of Fowls*—is Nature itself, which Spenser personifies as a bundle of contradictions. Though a "grand dame," Nature is hermaphroditic, self-generating, and at once young and old. Although (as shown in Canto 7) she is huge, much larger and taller than any of the other gods or powers in the universe, she is also hidden behind a wimpled veil, perhaps "To hide the terror of her uncouth hew": nature is very beautiful but also very terrible.[3] She is fittingly "Unseene of any, yet of all beheld" (Stanza 13), since humans experience nature's effects but not nature herself. Nature is also a picture of Christ, as Stanza 7 compares her to the Transfiguration, which the disciples behold in Matthew 17 as both indescribably wonderful and fearful. (Even Chaucer, Spenser says, "durst not with it mel [meddle]" and defers to ideas on the subject from Alan of Lille (Stanza 9).) Suggesting the superiority of nature over culture, Spenser writes that Nature's temporary throne, adorned with beautiful flora that bloom spontaneously around it, is far superior to that built by royal carpenters. Unbiased in her ecology, she "knittest each to each, as brother unto brother" (Stanza 14).

Such tensions are unknown to and unresolvable for Mutabilitie, a Titaness and personification of ever-present, universal change itself, who boastfully presents her case of sovereignty, and it is at this point that Spenser will resolve the question at hand: although we live in a world of decay and entropy, "Upon the pillours of Eternity,/That is contrary to Mutabilitie," subject to its own nature, God's creation rests in the eternal Sabbath. Through Mutabilitie, Spenser appropriates Pythagorean thought by way of Ovid (*Metamorphoses* XV) on *omnia mutantux*, "all things are changed."[4] In arguing that Change itself, and not Jove, is sovereign, Mutabilitie itemizes her control over creation: man, beasts, oceans and other bodies of water and the fish residing in them, the ancient elements of fire, air, water, and earth—"all are in one body, and as one appeare" (Stanza 25), and even the Seasons, Months, Day, Night, Hours, and Life and Death, as we see in the procession in Stanzas 28–46. In her closing remarks, Mutabilitie says, "all that moveth, doth mutation love" (Stanza 55), and Nature admits that this is true, though things change, she says (in a mix of Christian doctrine and Aristotle), to

> . . . worke their owne protection so by fate:
> Then over them Change doth not rule and raigne;
> But they raigne over change, and doe their states maintaine. (Stanza 58)

Some modern readers may recognize the notion, repeated several times in the canto, even by Nature, that all things change as a Darwinian fight for survival through the modification of species, but it is for Spenser much closer to the Aristotelian notion of "becoming." Even so, this extraordinary poetry cosmically decenters humanity: the notion that Man is sovereign is never even suggested in the poem, and humans in general are mentioned only along with the "beasts." Spenser upholds Christian hope for transcending chance and flux, but he also looks back to such Greek thinkers as Empedocles and Heraclitus in suggesting that we live not in a fixed world but in one that is centered largely on constant change. As John Watkins writes, Spenser's poetry "acknowledges eternity as the end of human striving, but it focuses primarily on the manifestation of eternal verities within the fallen temporal order" (94).

Spenser is one of many English poets to account for uncomfortable earthly truths and resolve them with a comforting conclusion, almost invariably in a Christian context. Tillyard writes that all Elizabethans, Spenser included, submitted to the same celestial order that underpins all their writing. They were all "obsessed by the fear of chaos and the fact of mutability." He cites Richard Hooker (*Of the Lawes of Ecclesiastical Politie*) as a case study. If nature, writes Hooker, "should intermit her course and leave altogether, though it were but for a while, the observation of her laws"—if the heavens and its celestial spheres, the sun, the moon, the seasons, winds, clouds, rain, and vegetation should quit working—"what would become of man himself, whom these things now do all serve?" (16). Similarly, Thomas Nashe's (1567–1601) "Litany in Time of Plague" addresses the brevity of human life, wealth, beauty, strength, and wit in strong imagery—"Fond are life's lustful joys,/Death proves them all but toys" (3–4)—though the poem ends positively for Christians of all social classes: "Heaven is our heritage,/Earth but a player's stage" (38–39). Shakespeare's Sonnet 64 recounts examples of "Time's fell hand" (1) having acted on "lofty towers" now "down-razed" (3); unlike many such sonnets of the era, there is no positive resolution. When the speaker thinks about Ruin, "which hath taught me thus to ruminate" (11), and the death of his love, he feels a sort of death himself and can only weep. Lorenzo states a similar notion to Jessica in *The Merchant of Venice* as he evokes the Pythagorean-Platonic music of the spheres: "Such harmony is in immortal souls,/But whilst this muddy vesture of decay/Doth grossly close it in, we cannot hear it" (5.1.63–65). Although it has been the subject of much scholarship over centuries, almost nothing is known of Shakespeare's personal religious beliefs, or even if he had any. Peter Ackroyd notes that he alludes often to Catholicism in his plays, yet he makes no religious declarations for himself. In his tragedies, "the religious imperative of piety and consolation are withheld; these are worlds with no god," though, Ackroyd acknowledges, this is a characteristic of Elizabethan drama in general (473–74).

Much of the poetry emerging in the first decades of the seventeenth century indicate a scientific depth inaccessible to Spenser or Shakespeare, even as most educated Elizabethans continued to have a somewhat vague conception of the universe as geocentric. A few years before Galileo published *Sidereus Nuncius*, which defends heliocentrism, Thomas Campion's song "What If a Day" (1606) functions as a sort of inverted *carpe diem* poem: seize the day because the universe is gigantic, the earth is small, time is fleeting, and (echoing Nashe) "All our joyes are but toyes" (9). The cosmic implication for humanity points, in this poem, to its irrelevance: "Earth's but a point to the world, and a man/Is but a point to the world's compare'd centure ["center"]" (13–14). George Herbert occasionally reflects the growing awareness of modern science that also connects with a *Scipio* image of the earth as a tiny point in heavens. In a letter to his ill mother, he writes, "As the Earth is but a point in respect of the heavens, so are earthly Troubles compar'd to heavenly Joyes" (372). In his "The Temper (1)," he feels unable to reach a God who has made a universe of "some forty heav'ns, or more" (5), but he resolves that such "distances belong to thee," God, who can "Make one place ev'ry where" (10, 28). Yet elsewhere, Herbert underscores the basic Christian anthropocentrism that was central to Renaissance thought. Man, he writes, "is ev'rything":

> For us the winds do blow,
> The earth doth rest, heav'n move, and fountains flow.
> Nothing we see, but means our good,
> As our delight, or as our treasure:
> The whole is, either our cupboard of food,
> Or cabinet of pleasure. ("Man" 7, 25–30)

As much as any English Renaissance poet, the new science emerges in the poetry of John Donne, though he was, like Herbert, not a scientific but religious writer. In his "At the Round Earth's Imagined Corners" ("Holy Sonnet 4"), the poet, aware of modern astronomy, feels the need to qualify Revelation 7:1, in which its author identifies "four angels standing at the four corners of the earth," as a trope, hence "imagined." (Similarly, Donne's contemporary Galileo, as in his famous *Letter to the Grand Duchess Christina*, warns against reading the Bible too literally.) One of Donne's richest poems from a scientific aspect is "The First Anniversary: An Anatomy of the World." The word "Anatomy," in this context, denotes a systematic analysis of a topic at hand. Man's corruption due to original sin is one of the poem's themes, as it is in many of Donne's poems. Does the poem suggest a nonanthropocentric view? Not if one looks at Donne's subsequent work, though Donne's interest in science and his appropriation of it suggests that it does. The same question arises in many writers from late antiquity to the present. Part of the woeful tone of the poem is related to the death of Elizabeth Drury, the daughter of Sir Robert Drury, one of Donne's patrons. The death of a young person is always tragic, and though the memories of most people fade quickly, Donne's poem has kept Elizabeth's name alive.[5]

Against the backdrop of the anniversary of a young girl's death, the speaker launches into a long, philosophical, ecclesiastical discussion of the place of humans in the universe—a universe grown larger with the new discoveries of Kepler and Galileo. Man, along with everything else in the sublunary region, is in a fallen state: "We are borne ruinous" (95). One of the poem's wittiest lines immolates wit itself in the context of ruin: "How witty's ruin?" (99). Of course, the answer to this (rhetorical) question is that ruin is not very witty. Many succeeding lines underscore the brevity of human life: "mankind decays so soon" (143); "This man, so great, that all that is, is his,/O what a trifle, and poor thing he is!" (169–70); "how poor a trifling thing man is" (184); "Be more than man, or thou'rt less then [sic] an ant" (190); and so on.

With line 205, Donne turns slightly to a new subject: how the new science, astronomy in particular, comes to bear on the human problem, as "new Philosophy calls all in doubt,/The element of fire is quite put out" (205–06). The "new Philosophy," the Copernican system, is the result of original sin—things are much more difficult because of the Fall. Line 206 is a pun marking the turn from Aristotelian/Ptolemaic (the elements of earth, water, fire, and air) science to a heliocentric one. Since this "is the world's condition now" (219), balance is gone, out of "proportion"—a word used twelve times in the second half of the poem. "Reward and punishment are beat away" (304), and now men seek "so many eccentric parts" (255). Anticipating the later romantic rejection of Newtonian science (e.g., Schiller's "The Gods of Greece" and Poe's "Sonnet—To Science"), man has now weaved a net and thrown it over the heavens (279–80), and we now "make heaven come to us" (282). The modern world is one of hubris, and so now the stars obey us (284). A series of lines, variations of which recur in the poem, underscore the depth of the poem's relative pessimism:

She, she, is dead; she's dead: when thou know'st this,
Thou know'st how wan a ghost this our world is:
And learn'st thus much by our anatomy,
That it should more affright, than pleasure thee . . . (369–72)

Donne would return rarely to the ideas stated in this remarkable poem, preferring in subsequent works toward more fully Christian subjects.

Topographical poetry dominated English letters in the early to mid-eighteenth century, and its focus on the British landscape is an important influence on the romantic explosion that occurred later in the century. Various arguments have been made for poets such as Thomas Warton and Thomas Gray as the first English romantics, and the case is persuasive to some degree. These and other poets held forward-looking ideas along with some old ones. Douglas Bush notes that both James Thomson (*The Seasons*) and Graveyard poet Edward Young ("Night Thoughts") veer toward vaguely evolutionary ideas, and both suggest man's limited vision (65, 67). A contemporary of these poets, John Dyer (1699–1757) is one of the more important eighteenth-century poets to write about nature, often within the context of ruins and often with a political perspective. In his pleasant topographical poem "Grongar Hill," the speaker watches rivers which are

Like human life to endless Sleep!
Thus is nature's vesture wrought,
To instruct our wand'ring thought;
Thus she dresses green and gay,
To disperse our cares away. (98–102)

The poem compares the courts, where one may search for but not likely find peace, unfavorably with the Welsh countryside. Resisting allegory, "Grongar Hill" underscores the primacy of physical nature itself (see Goldstein 29).

Dyer had no taste for the eighteenth-century fashion for ruins typified by Roman columns in gardens of noblemen (Louis XIV included such columns in the gardens of Versailles), but he visited Rome in 1724 and became fascinated quickly and deeply by its ruins. The fruit of this fascination, he published *The Ruins of Rome* in 1740. Dyer is aware of "the resistless theme, imperial Rome!" (15), which is now

Fall'n, fall'n, a silent heap; her heroes all
Sunk in their urns; behold the pride of pomp,
The throne of nations fall'n; obscur'd in dust;
Ev'n yet majestical: the solemn scene
Elates the soul . . . (16–20)

On a walking tour, not unlike "Grongar Hill," Dyer alludes to many places and historical events in history, and the spectacle of the place moves him, more than once, to think about England. The Roman patriots inspire the speaker's nationalism: "high ambitious thoughts inflame/Greatly to serve my country, distant land,/And build me virtuous flame" (127–30). He has the wish to have defenders of liberty, "Parent of happiness" (211), represented by Cicero during the Catiline conspiracy, to "be Britain's care" (213) and, more specifically, "To check the ravage of tyrannic sway;/To quell the proud; to spread the joys of peace" (225–26). Dyer turns to England once again at the end of the poem, warning his countrymen (as did Cicero and Seneca) against the complacency of luxury:

O Britons, O my countrymen, beware,
Gird, gird your hearts; the Romans once were free,
Were brave, were virtuous.—Tyranny howe'er
Deign'd to walk forth awhile in pageant state
And with licentious pleasures fed the rout . . . (511–15)

The earth is filled with the ruins of great kingdoms—Asshur, Cham, Elam, Greece, and Rome (542–45). By implication, England could be next. Hence Dyer finds a sort of lesson in the ruins of Rome to quell human pride, but not from the ruins themselves or their commonplace symbol of human temporality and proof against hubris but rather from the ancient history that unfolded in the area. Tim Fulford, contrasting Dyer with the nationalizing of the pastoral, writes that, unlike James Thomson and James Grainger, Dyer does not wish to glorify Britain; in "Grongar Hill," he sees the vanity of human ambitions (117). Yet Dyer is not explicitly antianthropocentric in his poetry. His late poem "The Fleece" is, in fact, directly anthropocentric in its celebration of the draining of a swampy fen—"A dreary pathless waste"—in the British countryside that is not beneficial to the British realm and unsuitable for human use, "Till one of that high-honoured patriot name,/Russel, arose, who drained the rushy fen," converting it "from chaos drear,/To raise the garden and the shady grove" (170–71, 175–76). Russel is Francis Russell, 4th Earl of Bedford (1593–1641), who led the drainage works the Fens of Cambridgeshire.[6]

A handful of eighteenth-century British poems veer, if slightly, away from such anthropocentrism. In Thomas Grey's celebrated "Elegy Written in a Country Churchyard," Fulford writes, "the poet does not sit above, but stands in, the landscape. He has not position of security . . . Nature, he sees, is too indifferent to human concerns to leave the viewer feeling powerful . . . nature has, in fact, its own rhythms of renewal, which grimly mock the finality of the grave" (118). The poem suggests a romantic outlook in its natural imagery as well as in the concluding autobiographical turn. Yet virtually all eighteenth-century English landscape poetry is anthropocentric. For James Thomson and William Cowper, "'God made the country,' and made it to bless man" (Fulford 125).

Although interest in ruins had been a poetic theme since the early Renaissance (and an artistic one for painters from then well into the nineteenth century), Enlightenment writers expressed an even deeper interest in the decline and fall of empires. Certainly, the chief work in this realm is Edward Gibbon's *The History of the Decline and Fall of the Roman Empire*. The final chapter of Gibbon's great work sums up the fall of Rome in one memorable sequence. The humanist Poggius and a friend stand on the Capitoline hill, where they

> reposed themselves among the ruins of columns and temples; and viewed from that commanding spot the wide and various prospect of desolation. The place and the object gave ample scope for moralising on the vicissitudes of fortune, which spares neither man nor the proudest of his works, which buries empires and cities in a common grave; and it was agreed, that in proportion to her former greatness, the fall of Rome was the more awful and deplorable. (1062)

According to J.G.A. Pocock, Gibbon thought the empire's fall was the result of "success and excess" but that "[n]o theory of human progress could be constructed which did not carry the negative implication that progress was at the same time decay" (148–49).

Interestingly, Giambattista Vico, in his imaginative reconstruction of early humans (*The New Science*, 1725), theorized that Rome, a city that fascinated him, would eventually be reclaimed figuratively and literally by forests and then by forest-people of the north (Harrison 12–13). Vico roughly echoes Thomas Burnet, the English clergyman whose *The Sacred Theory of the Earth* (written originally in Latin and published in complete form in English in 1690) divides history into three periods: an antediluvian, a postdiluvian, and a future one physically like the first (Glacken, *Traces* 407). For Burnet, the idea that the universe was made for man is absurd: "we must not, by any means, admit that all nature, and this great universe, was made only for the sake of man, the meanest of all intelligent creatures that we know of; nor that this little planet, where we sojourn for a few years, is the only habitable part of the universe" (352).

Much British poetry in the century, following Newtonian physics that began to reveal the expansiveness of the universe, developed further the idea of a world that is not human centered. Mary Midgley points out the paradox with the rise of science in general that applies to much British poetry in the eighteenth century. Science, she writes, "has always seemed a particularly bright jewel in the crown of the titular lord" (Kant's term for man as teleological end in *The Critique of Judgment*), but "it turns out also, paradoxically, to be an axe cutting away the floor under his throne" (107). Using the same metaphor, Douglas Bush notes that eighteenth-century British poets lowered the human throne a little, but the throne remained (73).[7] Pope's *An Essay on Man* is heavily Newtonian, and in the poem's attempt (updating Milton) to "vindicate the ways of God to Man" (1.16), he confirms the Chain of Being, with Man "Plac'd on this isthmus of a middle state" (2.3), and concludes Epistle I with a seeming tautology, attacked furiously by Voltaire in *Candide*, "WHATEVER IS, IS RIGHT" (1.294). In the Great Chain, sources of which lie in Plato, "man" may be in the "middle state" of the cosmic scale, but he is hierarchically at the forefront on the earth. Yet the poem does provide a picture of humans amidst a larger context that suggests the basis for a soft anthropocentrism.[8]

Pope's "Epistle to Burlington" addresses Richard Boyle, the 3rd Earl of Burlington but also pays respect to Lord Cobham (Richard Temple), who ordered nature in the ways that he would order England, by suggesting that nature is at man's beck and call, though, as Timothy Fulford notes, Pope knew better and is concerned less with showing the order of the universe than with contrasting the English estate, emblem of a consensual body politic, with the landscape of French absolutism. The poem is his way of justifying the ways of Lord Cobham to his subjects by way of writing about Cobham's garden (Fulford 110). Similarly, James Thomson's *The Seasons*, notes Fulford, "turns the viewing of landscape into a confirmation of the landed classes' right to power" (112). In the poem, "the laborer doesn't labor. He is a generalized 'happy Swain,' a decorative consumer in a scene in which Nature produces without his efforts" (113). Such a nationalizing of landscape would have American adherents in the writings of Francis Parkman and in some of the paintings of Thomas Cole (see Chap. 5).

European Romanticism

For one of the chief early French romantics, François-René, vicomte de Chateaubriand (1768–1848), ruins are figures of time and are important in two of his most significant works, *Memories from beyond the Tomb* and *The Genius of Christianity*. As a "fanciful ruin-fancier" (Ginsberg 359), he is interested in natural cycles of time and especially with human temporality: the relationship between ruins and humans is that of a mirror (Markarius 113). In *The Genius of Christianity* (1802), he lays out the attraction and ethical value of ruins in a number of key passages:

> All men take a secret delight in beholding ruins. This sentiment arises from the frailty of our nature, and a secret conformity between these destroyed monuments and the caducity of our own existence. We find moreover something consoling, to our littleness in observing that whole nations, and men once so renowned, could not live beyond the span allotted to our own obscurity. Ruins, therefore, produce a highly moral effect amid the scenery of nature . . . And why should not the works of men pass away, when the sun which shines upon them must one day fall from its exalted station in the heavens? He who placed it in the firmament is the only sovereign whose empire knows no decay. (466–67)

Byron would see himself as "a ruin amidst ruins" (*Childe* 4.25), but the near fatalism of his confession becomes a key religious point for Chateaubriand: "Man himself is but a decayed edifice, a wreck of sin and death; his lukewarm love, his wavering faith, his limited charity, his imperfect sentiments, his insufficient thoughts, his broken heart,—in short, all things about him,—are but ruins!" (468). Roman Catholicism provides Chateaubriand a solace rejected by Byron and Shelley.

Chateaubriand echoes many church fathers over the centuries, including Augustine, by emphasizing the insignificance of man against the natural sublimity of God: "I am nothing; I am only a simple, solitary wanderer, and often have I heard men of science disputing on the subject of a Supreme Being, without understanding them; but I have invariably remarked, that it is in the prospect of the sublime scenes of nature that this unknown Being manifests himself to the human heart." We are conscious of our insignificance, he goes on to write, "in the presence of the Infinite" (172). Gothic churches echo nature, and their more antique aspects bring us to "a reflection on the nothingness of man and the rapidity of life" (89). He is amazed by the "prodigious elevation" given to man by God, yet this is understood within the backdrop of our brief time on earth, during which "we should thus discover a resemblance between our fleeting days and the eternal duration of the Sovereign of the universe" (91, 92).

Chateaubriand wrote *The Genius of Christianity* within the context of the free thought and atheism of the French philosophes. Against the revolutionary tide, Chateaubriand's book attempts to show the continued centrality of Christianity in culture and art. Countering the "sophistry" of contemporary atheists, who would reduce everything in nature, humans included, to the material, Chateaubriand, like the physico-theologians of previous centuries, believes that one may understand final causes in nature. For Chateaubriand, God exists because animals are so well adapted to their ends—all is "secretly arranged" by Providence (144) and "deducible from the wonders of nature" (174). Songs of birds, for example, are "ordained so expressly for our ears" (148).

He is not, however, always doctrinaire in his writing. Medieval Christianity was one of the key factors in continental romanticism, and *Genius* was one of the first romantic tomes of its expression. Like his contemporary Goethe (also a theist), he itemizes the unity of nature, as "different parts of the universe exhibit the same wisdom that is so plainly expressed in the whole" (141). Like God, "man, created in his image, is likewise incomprehensible" (159). Where everything in nature is harmonious, "grief has its seat in the heart of man" (161).

He sometimes preaches *contemptus mundi*, as when he writes of the persecution of the first Christians "strengthening in them this disgust of the things of this life" (88). But Chateaubriand is in fact a nature lover. Though man is often out of step, Chateaubriand is concerned with showing the "harmonies of religion and of nature" (473). Anticipating John Muir, who would write similar words, he writes, "The forests were the first temples of the Divinity" (90), and he makes the occasional political-conservationist statement tied to human over-reach. Crocodiles in the New World (which Chateaubriand visited in 1791) have their territory as ordained by God, but when man, "the great destroyer," appears on the scene, the crocodiles must "resign the empire to us" (165).

Chateaubriand's contemporary and (arguably) equally important French Romantic, Anne Louise Germaine de Staël-Holstein (1766–1817), known commonly as Madame de Staël, was one of the most celebrated figures in the continent in her day as a writer, advocate for German literature, and as a chief opponent to Napoleon. Her novel *Corrine, or Italy* (1807) shows the mind of a humanist, a champion of free people, and an enduring model of a strong, independent woman. Some of the novel takes place in the ruins of Italy, especially Rome, the ruins of which Corrine sees as a symbol of man's greatness, his "divine spark" (65). Corrine tells her melancholy love interest, Oswald, "You know, my Lord, that the sight of the tombs, far from discouraging the living, was thought by the ancient Romans to inspire a new emulation, and so these tombs were sited on public roads; young people were thus reminded of famous men and silently invited to imitate them" (76).

Though Corrine projects optimism in the face of difficult times, she also acknowledges ruins as symbols for human transience and reminders of human limitations. Near Saint Peter's Basilica, she quotes Tasso: "cities fall, empires disappear, and man regrets his mortality!" (60). Observing the great sanctuary, she observes (not out of step with Chateaubriand) "how puny man is in the presence of religion" (61). She regards plants growing on the Palatine Hill and remarks to Oswald that "Nature has regained its empire" (68). She delivers a conservationist and ecological statement as the lovers complete their tour of Rome, which is deserted because of bad air:

> The lack of trees in the countryside round the town is probably one of the causes of the unhealthy air, and perhaps that is why the ancient Romans dedicated the woods to goddesses; they wanted to make the people respect them. Now, innumerable forests have been cut down. In our day could there be places sacred enough not to be laid waste by greed? (86)

Corrine travels to Naples and visits Mount Vesuvius and discovers that its slopes are not hospitable to humans. In the town of Pompeii she finds "ruins upon ruins and tombs upon tombs" (199).

Pompeii would be a popular subject for many other writers in the early-nineteenth and mid-nineteenth century. In his celebrated poem "La Ginestra" ("The Broom Plant"), the great Italian poet Giacomo Leopardi issues a forceful attack on human-centeredness: Mount Vesuvius is "cruel" and a "destroyer," indifferent to human life, as is the entire universe. (The theme of a world without people recurs in many of Leopardi's works, including some of his *Operette Morali.*) Pompeii was also a source of reflections for Edward Bulwer-Lytton, whose *The Last Days of Pompeii* (1843) was extremely popular in its day, for Gerard de Nerval in his novella *Le Temple d'Isis: Souvenir de Pompeii* (1843), and Théophile Gautier with his Hoffmann-inspired short story "Arria Marcella" (1852), in which a young man encounters the spirit of a young woman imprisoned in the volcano's ashes. The important naturalist Alexander von Humboldt would witness the mountain's eruption in 1805 and climb it six times within the following few days (Wulf 143).

Humboldt's friend Goethe, too, was interested in Vesuvius, which he climbed in 1787 and discusses in *Italian Journey* (see 571–74). Goethe would be, among his many other contributions to literature and culture in general, a major writer on Italy. His aesthetic view of Rome, distinct from the historical view of others up to his day, would play a large role in romantic poetry and to ruins literature in its association with freedom and appreciation for decay (Frajlich 16). Although Goethe may be counted as a humanist, his artistic vision is so wide that he is, like a handful of other writers I discuss, including A. R. Ammons, capable of envisioning a world without or one not centered around humans. Such works include "The Metamorphosis of Plants," which steers almost wholly away from the human subject in an attempt to show the outer and inner workings of flora and suggesting a kind of consciousness.

In his most famous work, *Faust*, Mephistopheles is a mouthpiece for some of the ideas I discuss. In the initial meeting between Faust and Mephistopheles, the latter asserts his will to enact a global apocalypse, thereby giving readers a sense of his essential destructive (yet static) nature:

> I am the spirit of perpetual negation;
> And rightly so, for all things that exist
> Deserve to perish, and would not be missed—
> Much better it would be if nothing were
> Brought into being. Thus, what you men call
> Destruction, sin, evil in short, is all
> My sphere, the element I most prefer. (1338–42)

In particular, Mephistopheles is negative toward man: "Let foolish little human souls/Delude themselves that they are wholes" (1347–48). Faust eventually gives in to Mephistopheles, who, unable to act beyond divine will (or natural law), like Milton's Satan, can at best only enact his negativity minutely—"Failing that grand annihilation," says Faust, "You try it on a smaller scale" (1360–61). And regarding the earth, Faust says,

> This solid lump cannot be shaken—
> Storms, earthquakes, fire and flood assail the land
> And sea, yet firmly as before they stand!
> And as for that damned stuff, the brood of beasts and men,
> That too is indestructible, I've found . . . (1366–70)

Near the end of the phantasmagoric Act III of *Faust Part II*, Faust proclaims to Helen of Troy the pastoral beauty of an idealized Arcadia in what is, writes David Luke, an "impressive celebration of the eternally productive forces of nature which outlast all cultures" (*Faust* xlii). Jane K. Brown writes, regarding *Faust Part II*, that "History is no longer solely the real world in which the Ideal can be perceived in the Real, but, as a realm of successive illusions, is also the relentless destroyer of all human achievement" (99).

Although Goethe is one of the greatest of humanists, he rejects humanity as the teleological end of creation. In his short dialogue poem "True Enough: To the Physicist," the speaker responds to the falsity he perceives in the lecture of a physicist, who is prone to think that "in every place/We're at the centre"; the speaker asserts, "Nature has neither core/Nor outer rind,/Being all things at once" (116). In *Conversations with Eckermann*, an account of Goethe's thought in his final decade, Goethe praises the book of a young unnamed scientist and pardons him for his "teleological tendency":

> "It is natural to man . . . to regard himself as the final cause of creation, and to consider all other things merely in relation to himself so far as they are of use to him. He makes himself master of the vegetable and animal world; and, while he claims other creatures as a fitting diet, he acknowledges his God, and praises His goodness in this paternal care. He takes milk from the cow, honey from the bee, wool from the sheep; and while he gives these things a purpose which is useful to himself, he believes that they were made on that account. Nay, he cannot conceive that even the smallest herb was not made for him; and if he has not yet ascertained its utility, he believes that he may discover it in future." (314)

English Romanticism

The literary decentering of humans becomes more complicated in confronting the writings of the English Romantics, who were, at once, focused on external nature but also concerned centrally with human consciousness and individuality. I focus in this section on Lord Byron and Percy Shelley because they veer closest to my subject, though other chief writers of the period have much to add to the discussion as well. Across Europe and, later, in the United States, all romantics, by definition, rejected the mechanical order initiated by Newtonian science. Where the rationalists of the eighteenth century saw the universe as mechanical and inorganic, the German idealist Friedrich Wilhelm Joseph von Schelling (representing the views of many, perhaps most, romantics of all nationalities) believed that nature is an organism that achieves consciousness through humans (M. Ferber 86).

In my first chapter I discuss a few writers who assert that anthropocentrism, somewhat ironically, became a more dominant mindset *after* the Copernican revolution, and there is much evidence for this notion. The romantic fascination with the Sublime combined with new scientific interests that replaced nature as a reflection of God, Ashton Nichols notes, "inaugurates a new brand of naturalistic anthropocentrism, the consequences of which are still with us in many ways" (8). Nichols is correct, yet the application of a neat binary is, I believe, problematic. The writings of the romantics demonstrate that these seemingly contrasting ideas are not mutually exclusive, just as the Enlightenment both highlighted human reason but also exposed the tininess of humans in the universe. In contrast to the distance with which earlier English nature writers such as Gilbert White perceive natural objects, Raymond Williams writes that the "separation" in the new green verse of William Wordsworth and John Clare is "mediated by a projection of personal feeling into a subjectively particularized and objectively generalized Nature" (134). Romantic poets elevated the self and human consciousness even as they also elevated external nature in itself—birds, trees, flowers, and other natural entities.

Literary critics have argued that the romantic revolution is something of a misnomer and, in fact, the result of a gradual shift away from rationalism to emotion reflected in many eighteenth-century poems.[9] As much as it is a revolutionary movement, romanticism may be understood as the culmination of earlier ideas we have seen in the French philosophes and some eighteenth-century English poets in which writers envision a world outside of humans, from the point of view of the nonhuman, and, sometimes, a world without humans. Thoughts on the Sublime—e.g., tall, rocky cliffs that overwhelm and underscore the feebleness of mere human powers—persist and develop with the romantics. In his *Critique of the Power of Judgment*, Kant writes that overhanging, threatening rocks, thunderclouds, volcanoes, high waterfalls, and the like "make our power of resistance of trifling moment in comparison with their might" (110). Such thoughts also apply to ruins, ranging back to Francis Colonna, meditations upon which produce both wonder and feelings about the smallness and transience of human life.

The melding of external nature and inner consciousness culminated in Wordsworth, for whom nature involves community both in itself and in its relationship to humans. (Much the same could be said of Thoreau, who, a generation or so later, would write in his journal, "What is Nature unless there is an eventful human life passing within her?" (*Journal* 5: 472).) *Lyrical Ballads* is of course a revolutionary volume of poetry, a repository of "the spontaneous overflow of powerful feelings," as Wordsworth defines "good poetry" (448), a book in praise of rusticity and simplicity, of the vernacular as opposed to what Wordsworth and Coleridge saw as the stilted neoclassical language of Erasmus Darwin and others. Yet the book is also revolutionary in its views on the natural world and the human relationship to and membership in it. Romantic reflections on the innate connection of humans with nonhuman nature, replete with the "I" (*The Prelude* in particular), may give the appearance of anthropocentrism (again, as it often seems with Thoreau and Emerson). But if the human mind is a part of nature, as it is for most of the chief romantics, then the "I" speaks not out of a self-aggrandizing will to own or wield preeminence but as an expression of kinship and sympathy for nonhuman nature. In "Tintern Abbey," Wordsworth places this kinship on the forefront when he writes of a "sense sublime,"

> Whose dwelling is the light of setting suns,
> And the round ocean and the living air,
> And the blue sky, and in the mind of man . . . (97–99)

Thoreau puts the idea directly in a letter he wrote as a young man in which he says he dreams "to be nature looking into nature with such easy sympathy as the blue-eyed grass looks in the face of the sky" (*Correspondences* 45). The poet, Wordsworth writes in the famous Preface, "considers man and nature as essentially adapted to each other, and the mind of man as naturally the mirror of the fairest and most interesting properties of nature" (455). The imagination, he writes in "Tintern Abbey," helps us "see into the life of things" (49).

Other poems in *Lyrical Ballads* explore the connection of external nature and the human mind. In Coleridge's *The Rime of the Ancient Mariner*, which may be read as an ecological allegory, the Mariner forsakes the unity of God's creation, murders the albatross (humanized by Coleridge as a worshipper of God), and after a trying journey, finds grace. "The Foster-Mother's Tale," also by Coleridge, concerns a baby found next to a tree wrapped in mosses. As he grows, he has no taste for institutional religion, and instead realizes religion through realizing creation, "as he was a bird himself" (Wordsworth, *Lyrical* 17). In Wordsworth's "Lines (Left upon a Seat in a Yew-Tree)," Ralph Pite writes, "the mind and world replicate each other because they are kin to one another," yet the poem also "engenders a temperate sense of one's own relative insignificance" (184, 188). The young man of the poem leaves the corrupt town for the solitude of nature. In the wilds, he feels a connection with the world impossible in the city: "The world, and man himself, appeared a scene/Of kindred loveliness" (*Lyrical* 40–41). Human pride, he feels, "Is littleness" (51); nature, meanwhile, is pride's contrary, and it teaches wisdom to all living things and "leads to love" (59). The poem works both as an aesthetic statement for simplicity, but it also functions ethically and as a counter to anthropocentrism.

Our consciousness of our essential relations to the natural world, or the lack of them, Wordsworth suggests, holds strong implications for our ethical treatment of nonhuman nature as well as our fellow humans. This is central in "The World Is Too Much with Us," which claims that "we lay waste our powers:/Little we see in Nature that is ours" (2–3)—the natural base that sustains us all, whether we acknowledge it or not. The speaker would rather be a pagan than a vulgar Christian; unlike most modern Christians, the pagan is aware of the vital human connection with nature. In "Lines Written in Early Spring," the ethical idea is even more up front. It occurs to the speaker as he sits in a grove observing birds, flowers, and other natural phenomena, that he is a part of these, and they are a link that runs through his own soul. Awareness that modern man has lost his connection to nature leads him to conclude, "Have I not reason to lament/What man has made of man?" (23–24).

However and whenever the romantic turn occurred, the melding of inner consciousness with external nature had arrived, before Wordsworth, by the late eighteenth century in the poetry of Anna Letitia Barbauld, Charlotte Smith, and Robert Burns. Barbauld's "The Mouse's Petition" shows compassion for a small rodent, and her poem "The Caterpillar" apostrophizes an even smaller creature with seeming genuine consideration for its own nature. Smith's long, posthumously published poem *Beachy Head* is set in southernmost Sussex but extends to vast geologic time as she considers the human history of the area in the context of an old earth. Man is not immortal, though his belligerence over the ages suggests he thinks he is:

Hither, Ambition come!
Come and behold the nothingness of all
For which you carry thro' the oppressed Earth,
War, and its train of horrors—see where tread
The innumerous hoofs of flocks above the works
By which the warrior sought to register
His glory, and immortalize his name . . . (419–25)

Like Barbauld, Robert Burns's "To a Mouse" (1785) apologizes to a rodent for having upset its life—the poem's subtitle is "On turning her up in her Nest with the Plough, November, 1785." The poem opens with what could be called apostrophe, but the personalization of the address is much closer to being the expression of connection with what the poem terms a "fellow-mortal":

WEE, sleeket, cowran, tim'rous beastie,
O, what a panic's in thy breastie!
Thou need na start awa sae hasty,
Wi' bickering brattle!
I wad be laith to rin an' chase thee,
Wi' murdering pattle!

I'm truly sorry Man's dominion
Has broken Nature's social union,
An' justifies that ill opinion,
Which makes thee startle,
At me, thy poor, earth-born companion,
An' fellow-mortal! (1–12)

Like Burns a peasant poet, John Clare shows as direct attention to particulars in nature as any poet before him, especially in his bird and animal poems. Nonanthropocentric poems such as "The Nightingale's Nest," "To the Snipe," and "Cowper Green" praise brambly, marshy lands that are *not* of practical or commercial use for humans but which are perfectly fitting for the birds and animals that inhabit them. The power and poignancy of these poems are stronger given Clare's personal conflicts with government enclosures and his mental health problems later in life. His "Elegy on the Ruins of Pickworth" (1818) was, he wrote, the result of helping "to dig the hole for a lime-kiln, where the many fragments of mortality and perished ruins inspired me with thoughts of other times, and warmed me into song" (qtd. in E. Robinson xvii). At the place of work ("'The Old Foundations' still the spot"), "the nettle grows/In triumph o'er each heap that swells the ground" (3, 5–6). One person owns half the land while others, "Like me but labour for support in vain" (20); like Seneca, he castigates "Luxury surfeit with excess" (21) (though unlike Seneca, Clare was poor). He observes that "fragments of mortality abound" (40) and imagines that there is "not a foot of ground we daily tread . . . But holds some fragment of the human dead" (42, 44).[10] Similarly, his "The Lament of Swordy Well" is an extended prosopopoeia representing the ancient stone quarry first used by the ancient Romans, though the lament is also that of peasants who worked the land before enclosure. Now private property, the quarry has been diverted from its age-old, largely nonanthropocentric purpose of caring for birds, bees, butterflies, and rabbits: "For gain has put me in a pound/I scarce can keep alive" (151–52). The quarry's new *telos* having dwindled to mere profit, soon "My name will quickly be the whole,/That's left of Swordy Well" (207–08).

Fascination with ancient ruins (seen in Smith's *Beachy Head*) carried over from the neoclassical eighteenth century into the romantic nineteenth century and became a central idea among English poets, especially Byron and Shelley, who use it more directly as an antianthropocentric trope, though I do not suggest either poet's viewpoint is consciously nonanthropocentric. Aesthetically, interest in ruins is linked to the romantic notion that poetry is defined as much by absence as by presence. Friedrich von Schlegel is central to this notion. "The ruin," he writes, "conjures up absence," an idea he applies to poetry: "Similar to a work of art in miniature, a fragment must completely detach itself from the environing world and, like a hedgehog, close in on itself" (qtd. in Makarius 147).

Poetry, he writes, "should forever be becoming and never be perfected" (qtd. in J. Hillis Miller 25). If the romantic believes that there is no structure to things, then he or she is comfortable in a ruined world without civilization because it more closely represents the underlying (non)structure or fragmented reality of existence with more fidelity than the ordered city. Not coincidentally, Byron and Shelley were particularly interested in ruins, which are, by definition, fragments.

Lord Byron

Isaiah Berlin argues that although the genesis of Romanticism occurred in Germany, it perhaps found "its most passionate expression" in Lord Byron, "the leader of the entire romantic movement" by his embodiment of the two key romantic values: "the will and the absence of a structure of the world to which one must adjust oneself" (131–33). Byron was well acquainted with topographical poetry from Virgil to Wordsworth and was also well aware of poetry about ruins, but, Robert F. Gleckner notes, it is not until the fourth and final canto of his long topographical poem *Childe Harold's Pilgrimage* (1812, 1818; hereafter *CHP*) that Byron "settles the poem squarely in that tradition" (269). Byron is not as concerned with history or nature as he is with human mortality. Thus Byron, tourist of European ruins, fleeing the spotlight of England, sees himself in the canto as "a ruin amidst ruins" among "Fall'n states and buried greatness" (4.219–20). Gleckner contends that Rome is, for the Byron of Canto IV, a symbol for man, "a microcosmic Rome" (280). Byron's focus on human consciousness is one reason the poet is seldom included in contemporary ecocritical studies. But his love of nature combined with a questioning of anthropocentrism, often within the context of ruins, makes his work worthy of reexamination within an ecocritical frame.[11]

Byron had been familiar with ruins from the near beginning of his life at Newstead Abbey, and he visited and lived near ruined sites in Greece and Rome, which was for Byron a "marble wilderness" (*CHP* 4.710). Byron would write in an 1816 letter, "I have been familiar with ruins too long to dislike desolation" (qtd. Marchand 259). For Byron, the heavenly consolations evoked in ruins poetry in *The Wanderer*, Spenser's *Complaints*, and Gray's "Elegy Written in a Country Churchyard" are illusory. Although he was a nature and animal lover, these did not provide him a lasting, Wordsworthian solace. In this sense, Byron had a more forward-looking, modernist view of ruins and the temporality of man than previous writers, including Wordsworth. Gleckner writes that the Byron of *CHP* IV has more in common with Eliot, Yeats, Joyce, and Faulkner than with Spenser (271). In such poems, Douglas Bush writes, Byron shows "the littleness of man and the nothingness of life" (100).

More than any other writer, Lord Byron established the romantic taste for ruins. In its admiration for wild nature on the one hand and its speculations on the brevity of human civilizations on the other, *CHP* is in many ways traditional, but it also prefigures ideas explored in detail in more modern works. The third of the poem's four cantos is especially important in the development of the idea I have been discussing. The first two cantos made Byron famous, but they are often overlooked today in favor of the last two. Yet even early in Canto I, the speaker Harold notes that in Portugal "the goodly prospects o'er the hills expand!/But man would mar them with an impious hand" (210–11). And he suggests the brevity of human civilization, developed more fully in Canto III, by asking rhetorically of a Spanish vista, "See how the Mighty shrink into a song!/Can Volume, Pillar, Piles preserve thee great?" (401–02). In Canto II, he indicates, here and there, a Wordsworthian desire to merge with untamed nature (see Stanza 25), and he sometimes combines the expression of his love for wild nature with descriptions of beautiful women (see Stanza 37).

Byron wrote Canto III in the wake of his failed marriage with Annabella Milbanke and the public response to it (as well as a response to Napoleonic Wars), and this accounts partially with his need to escape England amidst the controversy. Harold (and Byron) needs to wander in wild nature because "in Man's dwellings he became a thing/Restless and worn, and stern and wearisome" (127–28). He opens the Canto by identifying Harold as a wanderer in nature—first a weed sailing in the ocean, then as a windblown cloud (16–17, 22–23). Echoing Wordsworth in the Lake District, the mountains are Harold's "friends" (109), while

> The desart, forest, cavern, breaker's foam,
> Were unto him companionship; they spake
> A mutual language, clearer than the tome
> Of his land's tongue, which he would oft forsake
> For Nature's pages glass'd by sunbeams on the lake. (113–17)

He proceeds in the Canto to call the Alps "the palaces of Nature," and though the mountains provide a feeling of sublimity—they expand but also appall the human spirit (596)—they are cold, inhuman, and, compared to fleeting human life, have "throned Eternity in icy halls" (593). Harold senses too much of the human hand even in nature, but on Lake Leman (i.e., Lake Geneva or Lac Léman), in Rousseau country, the Wordsworthian outer nature as source for inner nourishment reaches something of a culmination in Cantos 67–75. There is "too much of man here" (648), but he makes it clear that he is no mere misanthrope ("To fly from, need not be to hate, mankind" (653), echoing not Molière's Alceste but Philinte), but he suggests an ecocentric view in these Cantos: "Is it not better, then, to be alone,/And love Earth only for its earthly sake?" (671–72), and he bridges a Wordsworthian desire for self-realization in nature with an almost Emersonian, spiritual merging with nature:

> I live not in myself, but I become
> Portion of that around me; and to me,
> High mountains are a feeling, but the hum
> Of human cities torture . . . (680–83)

A few stanzas later, he asks, "Are the mountains, waves, and skies, a part/Of me and of my soul, as I of them?" (707–08). Byron affirms this view much less than Wordsworth does; for Byron, the internal human struggle is always preeminent. He often contradicts this position, as in his 1814 letter to Annabella Milbanke, who was an orthodox Christian: why he was born, "I know not—where I shall go it is useless to enquire. In the midst of myriads of the living & the dead worlds—stars, systems, infinity—why should I be anxious about them?" (qtd. in Marchand 164). Nevertheless, combined with his aesthetic and spiritual love for nature, his belief that man has meddled too much with it, and his need for nature as a place to escape "Man's dwellings," Byron suggests the preeminence of nature over human civilizations, its immutability versus the petty short-term projects of human empires. Nature is not an option for man—it cannot provide lasting comfort—yet Canto IV concludes with a Wordsworthian vow to "love not Man the less, but Nature more" (Stanza 178)—an idea that would be, more than a century later, common in the poetry of Robinson Jeffers, e.g., "be in nothing so moderate as in love of man" ("Shine, Perishing Republic" 9). *CHP* IV is largely about the past, famous places and figures, but on the lake he pauses to note that "the earth/Forgets her empires with a just decay" (636–37).

Canto IV, though less about nature, resumes this theme as Harold—now more than ever closer to Byron's own voice—tours Venice, where he notes that "States fall, arts fade—but Nature doth not die" (24). Moving to Rome, he observes a variety of sites, which, once great, now

> . . . bows her to the storm,
> In the same dust and blackness, and we pass
> The skeleton of her Titanic form,
> Wrecks of another world, whose ashes still are warm. (411–14)

The power of the poetic vision can create the past glories of Rome (see Stanza 104), but nature has reclaimed the ruins, as at a shrine at Palatine (the domus of Augustus, Tiberius, Caligula, and Nero which perished with the burning of Rome), where

> . . . hillocks heap'd
> On what were chambers, arch crush'd, column strown
> In fragments, chok'd up vaults, and frescos steep'd
> In subterranean damps . . . (956–59)

And finally it is not Imperial Rome that Harold confronts

> Whose arch or pillar meets me in the face,
> Titus or Trajan's? No—'tis that of Time:
> Triumph, arch, pillar, all he doth displace
> Scoffing . . . (986–88)

Harold's tour of the ruins leads, at the end of Canto IV, to what is apparently an autobiographical reflection, leading to the well-known apostrophe to the sea that concludes the poem. "Upon the shores, empires wash away, and their decay/Has dried up realms to desarts" (1634–35), but the ocean is "Unchangeable save to thy wild waves' play" (1636). Gleckner states that ruins for Byron represent "the hell of human existence" (76), but Jerome McGann, for one, disagrees with this assessment.[12] Byron wrote to his friend John Hobhouse that *CHP* IV "is an augury or prophecy for England to heed, or else, like Greece and Italy, it too will have its reward, and at no very distant period" (qtd. in Gleckner 268). But it is a vision that applies individually as well as nationally. As for contemporary England, so is it for historical man: through kingdoms, "From Jove to Jesus . . . man plods/His way through thorns to ashes" (1308, 1310–11). At the Temple of Diana, "columns strew the wilderness" where now jackals and hyenas dwell (1373–74).

Byron wrote his Faustian closet drama (or, as he called it, "metaphysical drama") *Manfred* between the period of his composition of parts III and IV of *CHP*, and it shares a similar gloomy outlook, an uneasy praise of nature, and ruminations among ruins. Manfred, perhaps less a representation of Byron than is Childe Harold, lives amidst nature in the Bernese Alps, preferring the wilds of the Jungfrau to the city. Although he takes "refuge" in the Earth's mysteries, "they can nothing aid me" (2.2.42). Humans, Manfred states, "name ourselves [nature's] sovereign," but we are "Half dust, half deity" (1.2.39–40). The inhospitable Jungfrau is beautiful to him ("sight of loveliness" and "sweet solitude" (2.2.9–10); later it is a familiar face to him (3.4.3–5)), but it is not wholly beneficent. Anticipating Thoreau's terrible prosopopoeia on Ktaadn, Manfred understands that nature is often destructive, even malignant, and capable of killing innocents in avalanches (284–85).[13]

Like Childe Harold, Manfred cannot find inner peace in nature, as lovely as it is to him; it is for external use only (see Manfred's apostrophe to the Earth in 1.2.7–12). Echoing the ruins of *CHP*, Manfred concludes that man's days are "carcasses and wrecks . . . bitterness" (2.1.57–58). In a celebrated soliloquy near the conclusion of the play, Manfred recalls his wanderings through the ruins of the Roman Colosseum, among "the chief relics of almighty Rome," where trees have grown "along the broken arches" and the stars shine through the "rents of ruin"—the final victory of nonhuman nature over human works (3.4.11–14). Even mighty Caesar's chambers now must "grovel" in "decay" (3.4.30). All this is important as Manfred concludes the play: when he dies, nature will remain, including the moon that has witnessed the rise, fall, and decay of Rome. He admires nature, but the bitter pill he must swallow is the fact that he can't take it with him, nor can it save him or provide him with purpose for his life and his loss of his beloved Astarte. "The mind . . . when stripped of this mortality, derives/No colour from the fleeting things without"; it is "Born from the knowledge of its own desert" (3.4.129, 133–134), and when Manfred dies, underscoring the impermanence of humanity and our lack of connectedness with the earth, the abbot concludes that Manfred's soul has "ta'en its earthless flight" (3.4.152).

During this period, Byron also composed "Darkness," a dream poem, or nightmare, about an environmental disaster of the first degree, though, as Byron writes, it is "not all a dream" (1). The poem suggests a narrative, but it tends more toward an extended description, and a totally black one. The sun having burnt itself out, "the icy earth/Swung blind and blackening in the moonless air" (4–5). Forests are burned up for light, which is the greatest commodity, until all the trees have disappeared (19–20). The sky is "dull . . . The pall of a past world" (29–30); there is still a sun, but it has been "banished."[14] Not just men but animals too begin to change their habits. Birds' wings are useless, and "the wildest brutes/Came tame and tremulous" (34–35). No love remains, only death, including cannibalism and an eternal war (the poem was written during the Napoleonic era) between the two remaining cities. Byron echoes Genesis 1:2:

> . . . The world was void,
> The populous and the powerful—was a lump,
> Seasonless, herbless, treeless, manless, lifeless—
> A lump of death—a chaos of hard clay.
> The rivers, lakes, and ocean all stood still,
> And nothing stirred within their silent depths . . . (69–74)

As in a backwards Genesis, after a time, everything has unraveled, and all that remains is the darkness, which has become the whole universe. Leslie Marchand points out that the only altruistic behavior in Byron's poem is performed by a dog (246).

Byron wrote "Darkness" during the 1816 Year without a Summer, which was the result of volcanic activity, especially Mt. Tambora, east of Java, though other eruptions, mostly in the East, also contributed to the pall of the sky, an end to photophosphorylation, and organic failure, including crops, across northern Europe.[15] The volcanic smoke of Tambora also inspired the writing of Mary Shelley's *Frankenstein* and a number of paintings by J.M.W. Turner—e.g., *The Lake, Petworth Sunset, Fighting Bucks* (c. 1829). Byron's poem would influence heavily a number of writers, including the "Last Man" poems of Thomas Campbell and Thomas Hood as well as Poe's "The City in the Sea."[16]

In Byron's three-act poetic drama *Cain: A Mystery* (1821), the moody, rebellious title character is visited by Lucifer, who, somewhat reminiscent of both Milton's Eve confronted by Satan and Goethe's Faust interacting with Mephistopheles, becomes the willing student of the fallen angel, who confirms Cain in his rebelliousness toward God. Lucifer takes Cain deep into space where, invoking Cicero's *Scipio* and the medieval model of the universe, he looks back and recognizes earth, from this viewpoint a "small blue circle, swinging in far ether,/With an inferior circlet near it still" (2.29–30). Lucifer enforces Cain's growing negativism toward his "dull earth," built out of dust (2.46); the rebellious Cain and the rebellious angel speak the same language. Cain is something of a nature lover (see 2.2.255–69), but, as with Manfred, nature is not enough to sustain him, to give him hope against oblivion. His tour of the universe makes him see the smallness of man, but, not humbling him, it turns him into a nihilist capable of murdering his own brother.

Percy Shelley

Many of the themes of ruins and the smallness of humans established in classical literature are echoed in the writings of Percy Bysshe Shelley, though he also gives humans—idealized and governed by Reason and Nature in his secularized millennial rhetoric—a much larger and radical role than almost any English poet before him (see Paley, Chap. 5). For Shelley, especially as a young man, his opposition to the idea of a human-centered universe, increasingly well grounded in the science of the day, is the basis for his secular apocalyptic rhetoric. The vastness of the universe—realized by Newton but made much vaster through the astronomical discoveries of William Herschel—highlights both the smallness of humans as well as the evidence for atheism that would appear in works such as *Queen Mab* (1813), Shelley's first extended work of poetry (Holmes, *The Age* 390–93). He wrote the poem in nine cantos while still a young man during a period of political activism. Highly ideological and radical, the poem recalls Lucretius in its didacticism, and it was inspired by a host of thinkers, including Hume, Godwin, Wollstonecraft, Rousseau, and the French writer Count Constantin de Volney.

Volney was a member of a sensualist group of thinkers extant during the French Revolution, the Idealogues, who argued for the emancipation of humanity from unenlightened religion. His book *The Ruins or Meditation on the Revolutions of Empires and the Law of Nature* (published 1791; translated 1795) was read and admired by many freethinkers, including Thomas Jefferson, who met Volney and translated a part of *The Ruins* into English, and the book's influence lasted well into the nineteenth century. For Volney, the spectacle of ruins is a fecund source for "the sublimest charms of contemplation and sentiment" (11). Contemplation of the ruins of Palmyra (which Volney never visited) leads him to the conclusion—reminiscent of Edward Gibbon on the Roman Empire—that the decline of earlier civilizations was the result of corrupted kings and priests as well as the neglect of natural law, ignorance, greed, and "self-love" (44). A major theme throughout the work is nature's reclamation and eclipsing of the glories of human works. Nature, he writes, seems to say to man, "I owe thee nothing, and I give thee life; the world wherein I placed thee was not made for thee," though nature allows man use of it (40). Volney applies the words "ignorance" and "cupidity" with regard to man repeatedly in the work; these are, he writes, "the twin sources of all that torments the existence of man!" They are "the evil genii that have wasted the earth!" They have "converted the splendor of a populous city into a solitude of mourning and ruins!" (44).

Following Volney's rhetoric, which includes the formulae for human perfectibility, *Queen Mab* attempts to demonstrate that human morality is not based on religion. Rooted in Shakespeare and a sort of poetic Mother Goose, Queen Mab is an intermediary between the divine and the human.[17] As Ianthe (modeled on Shelley's first wife, Harriet) sleeps on her couch, Mab, a fairy queen, gives her spirit a tour of the universe in her magic chariot. Flying off into space provides a perspective, reminiscent of Cicero's *Scipio*, on the relative smallness of the earth—"The smallest light that twinkles in the heaven" (1.251). Mab points to a faraway earth and Ianthe recognizes her "kindred beings" in the thousands which seem from this perspective "like an anthill's citizens" (2.99, 101). A direct allusion to Volney, Mab points out "Palmyra's ruined palaces" (2.110). Egypt's pyramids now stand but will fall over time, and "where Athens, Rome, and Sparta stood,/There is moral desart now" (2.162–63).

Shelley's target is less anthropocentrism itself than a morality based on religion; he attacks Judaism as well as Christianity while praising pagans such as Cicero. Man, he writes, has a "brief and frail authority" (3.320), while, on the other hand, Nature is omnipotent, its soul

> That formed this world so beautiful, that spread
> Earth's lap with plenty, and life's smallest chord
> Strung to unchanging unison, that gave
> The happy birds their dwelling in the grove,
> That yielded to the wanderers of the deep
> The lovely silence of the unfathomed main,
> And filled the meanest worm that crawls in dust
> With spirit, thought, and love; on Man alone,
> Partial in causeless malice, wantonly
> Heaped ruin, vice, and slavery; his soul
> Blasted with withering curses; placed afar

> The meteor-happiness, that shuns his grasp,
> But serving on the frightful gulf to glare,
> Rent wide beneath his footsteps? (4.90–103)

Nature, the "all-sufficing Power," is not partial to humans, as he apostrophizes, "Because thou hast not human sense,/Because thou art not human mind" (6.197, 218–19).

A similar theme of the frailty of humans in contrast to the relative permanence of nature persists in many of Shelley's poems, including his sonnets. One notable example is "Ozymandias" (1818), which is probably the most famous of all English Romantic ruin poems. The poem dramatizes concisely the hubris of asserting the permanence of empire. Walking in the Egyptian desert, the speaker confronts a "trunkless head of stone" (2), a nearby disembodied "shattered visage" (4), and a pedestal which reads, "My name is Ozymandias, king of kings:/Look on my works, ye Mighty and despair!" (10–11). Though not a "nature poem," the sonnet concludes with a simple image which, contrasted with the pretentiousness of statue's claim, underscores the fact of the preeminence of nature and time (the more fitting bases for "despair"): surrounding the ruins, "boundless and bare/The lone and level sands stretch far away" (13–14). In "Mutability" (1816), Shelley picks up an idea from Spenser,

> We are as clouds that veil the midnight moon;
> How restlessly they speed, and gleam, and quiver,
> Streaking the darkness radiantly!—yet soon
> Night closes round, and they are lost for ever . . . (1–4)

Yet the poem concludes without the Spenserian hope of heaven: "Nought may endure but Mutability" (16).

The prose fragment "The Coliseum" employs a ruins theme differently. At Eastertime, a young woman and her blind, aged father sit on a fallen column in the Coliseum. A man, a non-Catholic, approaches and speaks to them, committing a *faux pas* by assuming the old man can see. The narrator judges Catholicism "[t]he most awful religion of the world," a means by which people "worship the creations of their own power" (*Shelley's Prose* 139). The man states that "the spectacle of these mighty ruins is more delightful than the mockeries of a superstition which destroyed them" (141). That is, Christianity put an end to the great classical age of art, learning and architecture. Concerning the singing of the wild pigeons visiting the site, the old man says the birds experience sensations, spirit, and reason of their own, independent of humans. Near the conclusion of the fragment, the daughter mentions death—a natural subject for an old man sitting among ruins—but the old man gives a rather long speech meant to persuade her that they should not talk about death.

Another prose piece, "Essay on the Devil and Devils," works more explicitly to unravel the idea of a human-centered universe by showing, logically but also with tongue in cheek, the unlikelihood of the Christian belief in Satan and Hell. How could such a potent foe be allowed to exist by an omnipotent deity? Does God allow Satan free reign in the vastness of creation? The essay draws from science of the day (e.g., Herschel) to disprove superstitions (e.g., that the sun is hell). The essay allows Shelley to stretch out in fuller parodic mode, anticipating some of the later writings of Mark Twain:

> The sphere of the operations of the Devil is difficult to determine. The late invention and improvement in telescopes has considerably enlarged the notions of men respecting the bounds of the Universe. It is discovered that the earth is a comparatively small globe in a system consisting of a multitude of others which roll round the Sun; and there is no reason to suppose but that all these are inhabited by organized and intelligent beings. The fixed stars are supposed to be suns, each of them the center of a system like ours. Those little whitish specks of light that are seen in a clear night are discovered to consist of a prodigious multitude of suns, each probably the center of a system of planets. The system of which our earth is a planet has been discovered to belong to one of those larger systems of suns, which when seen at a distance look like a whitish speck of light; and that lustrous streak called the Milky Way is found to be one of the extremities of the immense group of suns in which our system is placed. The heaven is covered with an incalculable number of these white specks, and the better the telescopes the more are discovered and the more distinctly the confusion of white light is resolved into stars. All this was not known during the gradual invention of the Christian mythology and was never suspected by those barbarians on the obscure extremities of the Roman Empire by whom it was first adopted. (*Shelley's Prose* 270–71)

In what many have argued is his greatest work, the lyrical drama *Prometheus Unbound* (1820), written in Italy, Shelley employs contemporary science in the framework for the human victory over tyranny, yet science is subservient to poetry, which is, he writes in *A Defence of Poetry*, "at once the centre and circumference of knowledge; it is that which comprehends all science, and that to which all science must be referred" (*Shelley's Poetry* 531). Prometheus of Greek myth is the champion of humankind who stole a spark of fire from heaven and taught humans the sciences in order to improve their harsh lives. A revision of Aeschylus' *Prometheus Bound* (and an inversion of Milton's *Paradise Lost*, a heavy influence in the poem), Shelley has Prometheus liberated by "alternative," benign forces as he is released while the repressive, supreme god Jupiter is vanquished. Without denying the central role of revolutionary man, some passages reflect the astronomy of Herschel in pointing to the relative smallness of the earth, as when Prometheus' mother, Earth, points out the "million worlds which burn and roll/Around us" (1.1.163–64). A complex work with various interpretations, some of the poem operates as versified passages from contemporary scientific writing (Chandler 347). Richard Holmes notes that Prometheus' new fire in the poem is inspired by recent experiments with electricity (*The Age* 506). But Shelley also refers to geological phenomena in the poem, as when a volcano erupts in Scene 2.4, and its "terrible shadow floats/Up from its throne, as may the lurid smoke/O earthquake-ruined cities o'er the sea" (150–52). The volcano will lead to the new restoration of Nature, its division healed, but it is also a political idea about an uprising and revolt against tyranny (see Holmes, *The Pursuit* 504–05). In Shelley's moral purpose of helping humans toward freedom from tyranny, including, as he sees it, freedom from oppressive religion, *Prometheus Unbound* is not antianthropocentric in its rhetoric, even as science, especially astronomy, pulls Shelley in that direction, in this work as in others.

The Last Man

The early part of the nineteenth century marks the development of the "last man" narrative, which would become, in relatively realistic fiction and in full-blown science fiction (and genres in between), a unique means for reflecting on the place of humans in the universe. The theme was established by Jean-Baptiste François Xavier Cousin De Grainville's *The Last Man* (1805), which is acknowledged widely as the first novel to depict the demise of the human race. More well-known is *The Last Man* (1826) by Percy Shelley's famous second wife, Mary Shelley. Though the novels, with identical titles, are obviously similar in subject—neither novel concerns, until their conclusions, the *very* last man—they vary widely in their rhetoric. In Grainville's novel, nature depends on humans for its existence, while in Shelley's, nature goes on without humans.

Grainville's novel reflects a central strain in English Romantic literature of widespread terror in the wake of the French Revolution and the Napoleonic Wars. The Terror probably shortened Grainville's life. He was a priest in opposition to the philosophes, but he took the requisite oath to the Republic, left the priesthood, married (unhappily), and died, likely by suicide, in 1805, and his novel was published posthumously. A deeply religious, didactic work, influenced by Milton and the book of Revelation, Grainville's novel is set in the far future. The deterministically named Omegarus, the son of the King of Europe and last born male, is persuaded by Ormus, the Spirit of Earth, to find a female with whom he may regenerate the population: "If you or she should die, the earth will disintegrate, sink back once more into chaos, and my life will be extinguished forever" (16). Omegarus locates the last fertile woman, Syderia, by flying to Brazil in an airship, but, after traveling back to Europe, the biblical Adam, the *first* man, forbids the couple from reproducing because God has decided to end humanity. Ormus (reflecting Christian orthodoxy) states that nature cannot exist without humans, falls into despair, since the planet cannot survive without humans. As he battles Death, the earth "shook, was blown out of orbit, and was torn asunder" (134).

Shelley's novel, by contrast, questions subtly the idea of man as the end of nature and suggests antianthropocentrism. Man is vulnerable, pretentious in the belief of his supremacy, demonstrated in the novel's second half, as the plague sweeps across Europe and, eventually, England. Shelley was, like her late husband, well informed on contemporary science, and her main character and narrator Lionel Verney writes,

> What are we, the inhabitants of this globe, least among the many that people infinite space? Our minds embrace infinity; the visible mechanism of our being is subject to merest accident. Day by day we are forced to believe this. He whom a scratch has disorganized, he who disappears from apparent life under the influence of the hostile agency at work around us, had the same powers as I—I also am subject to the same laws. In the face of all this we call ourselves lords of the creation, wielders of the elements, masters of life and death, and we allege in excuse of this arrogance, that though the individual is destroyed, man continues for ever.
>
> Thus, losing our identity, that of which we are chiefly conscious, we glory in the continuity of our species, and learn to regard death without terror. But

> when any whole nation becomes the victim of the destructive powers of exterior agents, then indeed man shrinks into insignificance, he feels his tenure of life insecure, his inheritance on earth cut off. (230)

Once man was "a favourite of the Creator," a little lower than angels and given dominion over nature. Now, however, Lionel writes, addressing humanity, you are forced to "give up all claim to your inheritance, all you can ever possess of it is the small cell which the dead require" (316). Man had "existed in twos and threes; man, the individual who might sleep, and wake, and perform the animal functions; but man, in himself weak, yet more powerful in congregated numbers than wind or ocean; man, the queller of the elements, the lord of created nature, the peer of demi-gods, existed no longer" (320).

As Lionel's fellow humans drop like flies from the plague, he steps back to wonder how the world will continue without man:

> Will the earth still keep her place among the planets; will she still journey with unmarked regularity round the sun; will the seasons change, the trees adorn themselves with leaves, and flowers shed their fragrance, in solitude? Will the mountains remain unmoved, and streams still keep a downward course towards the vast abyss; will the tides rise and fall, and the winds fan universal nature; will beasts pasture, birds fly, and fishes swim, when man, the lord, possessor, perceiver, and recorder of all these things, has passed away, as though he had never been? O, what mockery is this! Surely death is not death, and humanity is not extinct; but merely passed into other shapes, unsubjected to our perceptions. Death is a vast portal, an high road to life: let us hasten to pass; let us exist no more in this living death, but die that we may live! (413)

Even so, nature continues to console those willing to seek it; Lionel's group gathers in Chamonix, France, where, he notes, "Sublime grandeur of outward objects soothed our hapless hearts, and were in harmony with our desolation" (424), and, considering suicide, the sight of the Apennines helps Lionel rally: "wild natural scenery reminded me less acutely of my hopeless state of loneliness" (457). Lionel believes that, if a new set of people somehow repopulate the earth, "we, the lost race, would, in the relics left behind, present no contemptible exhibition of our powers to the new comers" (455). He observes untended cattle in a dell moving towards their watering hole, grass rustling in the breeze, and the woods mellowed by moonlight. Ruin is the result of human constructs: the Earth abides:

"Yes, this is the earth; there is no change—no ruin—no rent made in her verdurous expanse; she continues to wheel round and round, with alternate night and day, through the sky, though man is not her adorner or inhabitant" (459).

Much more successful artistically and commercially, Shelley's *Frankenstein, or The Modern Prometheus* (1818; 1831), also addresses human arrogance in its self-proclaimed supremacy over the rest of nature. Immediately before confronting the monster (educated, in part, by reading Volney), in the French Alps, Victor Frankenstein beholds the sublimity of the natural scenery and asks, "Alas! why does man boast of sensibilities superior to those apparent in the brute; it only renders them more necessary beings. If our impulses were confined to hunger, thirst, and desire, we might be nearly free; but now we are moved by every wind that blows," and he proceeds to quote Percy's "On Mutability" (75). Confirming these thoughts, Victor confronts the monster face to face for the first time a paragraph later. At points, the novel hints at a pessimistic view of humanity, following Swift and Voltaire. Rejected by humans for his outward hideousness, Victor asks, "Was man, indeed, at once so powerful, so virtuous, and magnificent, yet so vicious and base?" (95). The monster, as for Victor (and Byron and Shelley)—it is a novel filled with Doppelgängers—finds relief in the sublimity of nature, but only temporarily.

Until the 1920s, the common view held that the universe is made up of the Milky Way. Some speculated that the universe was much larger, including Immanuel Kant, who wrote that the universe is "infinite, whatever that might mean" (qtd. in Holmes, *The Age* 123). Although the size of the earth has been further contextualized and diminished since the findings of Edwin Hubble, as I have shown, many writers since antiquity have held some sense of the minuteness of our planet. In the British romantic era, which is dated often from the late eighteenth century until the early 1830s, this understanding intensified. Romanticism was a response to the sense of an orderly world mirroring Newtonian science. Yet the romantics did not reject science. With the discoveries of William Herschel and his sister Caroline, the earth became even smaller and less relevant within the growing cosmic context. Herschel's discovery of Uranus, alone, doubled the size of the solar system, and his later papers show an awareness of "deep time," which prompted discussion of its philosophical implications for humans (Holmes, *The Age* 101, 203). Perhaps Herschel would not have been surprised by Hubble's 1921 discovery that the universe is 100 billion times larger than previously conceived. The universe is not only not human centered, it is not even Milky Way centered (Holmes 205).

As Richard Holmes suggests, such knowledge gave poets and other writers of the era a renewed sense of uneasiness. Byron's "Darkness" reflects cosmological disaster hinted at in Herschel's late papers (383), and Shelley reasoned that the universe's vastness is an argument for atheism as seen in *Queen Mab*. Later writers responded to science in other ways. The popular science writer Mary Somerville's 1834 bestseller *On the Connexion of the Physical Sciences* maintains a generally pious tone, but, likely reflecting ideas distilled from her reading of William Herschel, she notes near the beginning of the book that the study of the heavens,

> while they ennoble the mind, at the same time inculcate humility, by showing that there is a barrier which no energy, mental or physical, can ever enable us to pass: that however profoundly we may penetrate the depths of space, there still remain innumerable systems, compared with which, those apparently so vast must dwindle into insignificance, or even become invisible; and that not only man, but the globe he inhabits nay,—the whole system of which it forms so small a part,—might be annihilated, and its extinction be unperceived in the immensity of creation. (4; see also Holmes, *The Age* 458)

Through the writings of the Romantics, religious and secular views of the apocalypse merged in the late eighteenth and early nineteenth centuries. Though there was, for Percy Shelley, always a Newtonian-Godwinian rationalism at core (Wordsworth would later reject Godwin), the secular writers and artists borrowed tropes and imagery and perhaps ideologies from the Bible. Later, in turn, religious artists such as Thomas Cole would borrow secular apocalyptic ideas from Byron and make them religious again.

The Victorian Era

The early Victorian Era marks, among other things, the apocalypse as the stuff for popular consumption and parody. "Last man" novels by Grainville and Mary Shelley underscore apocalypse as a widespread fear for romantics, but by the fourth decade of the century, the theme was prime for spoofing, though there were certainly very many "last man" narratives to come in verse, fiction, and, in the twentieth century, film, TV, and the Internet.

Thomas Hood is best known as a moral humorist concerned with social matters, especially the poor, and he is sometimes called the first Victorian. His light poem (on a dark matter) "The Last Man" (1826) takes place in the year 2001. All of humanity has disappeared due to a plague except two men, the narrator, who is by occupation a hangman, and a knave/beggar, who, happy to see someone else alive, says to the hangman, "Come, let us pledge each other,/For all the wide world is dead beside" (56–57). The plague becomes a social evener, as in Poe's "The Masque of the Red Death" and W.E.B. Du Bois's 1920 short story "The Comet": "rats leapt out" of masters' beds, and "the grandest palaces in the land/Were as free as workhouse sheds" (106–08). The poem turns with line 139, as the narrator, insulted by the beggar's finery, hangs him, leaving himself the last man, "To have my own will of all the earth" (171). Yet he feels unending guilt as well as consternation that for him there is no way out. He hesitates to hang himself, "[f]or there is not another man alive,/In the world, to pull my legs!" (221–22).

A young Charles Dickens would parody gently the moody attraction of apocalyptic literature in his first novel, *The Pickwick Papers* (1837), as at a gathering of friends and acquaintances, including Mr. Pickwick, an old clergyman delivers a poem he wrote as a younger man entitled "The Ivy Green," a part of which states, the ivy "creepeth o'er ruins old!/Whole ages have fled and their works decayed," nations have come and gone, but the ivy "shall never fade." After the clergyman recites his melancholy poem a second time so that a visitor may copy it down, the kindhearted, patient Mr. Pickwick tries to enliven the atmosphere, though without deflecting attention away from the clergyman, by asking him about his experiences in ministering the gospel. But his appeal is to no avail, as the old clergyman replies that his experiences "have been of a homely and ordinary nature" (72–73). Similarly, an 1846 *Punch* magazine article, "London in A.D. 2346," spoofs a presentation of a paper to the "New London Archaeological Institute" on the old city of London. Reports that officers of the institute "have been compelled to assign conjectural characters and destinations to the several buildings and works discovered . . . belonging to a 'bygone and barbarous race.' Conclusions include the attribution of remains of a 'low and singularly unsightly range of building' containing paintings 'crowded together in low close rooms,' to the National Gallery" (V. Zimmerman 135).

Richard D. Altick writes that the positivist response to Darwin in the late Victorian period was "Man Exalted instead of Man Degraded. Since no divine agency could be relied upon to ameliorate his condition, man must turn to himself to make whatever he can of his life. The only answer to his tragic destiny is found in his fully realizing and employing such powers as he possesses" (235–36). Of all nineteenth-century English poets, none was more concerned about the implications of modern science for poetry than Tennyson (Bush 109). Tennyson was influenced particularly by French natural historian Georges Cuvier, whose belief in the fixity of species was at odds with the organic evolution proposed by Maupertuis, Diderot, and Lamarck (Hankins 157). (The influence of Cuvier on literature and the arts in general was strong, spanning Balzac and, as discussed in Chap. 5, the painter Thomas Cole.) Yet Cuvier proved conclusively that the species may go extinct by way of natural catastrophic episodes. His theory once and for all destroyed the idea that the Great Chain of Being was established by God as immutable. (In the twentieth century, Robinson Jeffers would place humans near the bottom of the chain.) Tennyson also read Scottish geologist Charles Lyell's *Principles of Geology* (1830–1833), which theorized and popularized the idea of uniformitarianism, which in essence places the age of the earth at an advanced, deep time age, against the accepted idea that it is around 6000 years old. The impact of Lyell and Cuvier is strong in many of Tennyson's poems but nowhere more than what may be his greatest work, *In Memorium A.H.H.* (1833–1849, 1850), which the poet wrote in honor of his Cambridge friend Arthur Henry Hallam, who died at the tender age of twenty-two in 1833.

The poem, published before Darwin's *On the Origin of Species*, accepts what Darwin would call natural selection, yet Tennyson did not feel fully at ease with the theory, before or after reading Darwin, and he was not of the "Man Exalted" camp. Much of the poem's power lies in the poet's reflections on the tension between Christian faith and the new sciences. The final sections of the poem affirm Christianity, and, against evidence of an immense universe filled with change, profligacy, and strangeness, the speaker musters the idea that "we trust that somehow good/Will be the final goal of ill" (54.1–2). Faith wanes, at least temporarily, in subsequent stanzas, including the poem's most quoted, in Section 56, as feminized nature (a figure for current science) proclaims coldly, "A thousand types are gone;/ I care for nothing, all shall go" (3–4). Noble man, nature's last work, "trusted God was love indeed/And love Creation's final law—/Tho' Nature, red in tooth and claw/With ravine, shriek'd against his creed—" (13–16). Change and seeming chaos are not limited to humans; these occur slowly, patiently in the natural world. Section 123, reflecting Lyell, turns toward mutability in the physical world, as the poet apostrophizes the earth—"what changes hast thou seen!" (2). Even the hills "are shadows" that "flow/From form to form, and nothing stands;/They melt like mist, the solid lands,/Like clouds they shape themselves and go" (5–8).

Such Keatsian melancholy derived partly through reading about new scientific discoveries persists in many of Tennyson's poems. His poem "Vastness" indicates the gloomy mood of his later years. Reflecting on day-to-day politics as well as "this poor earth's history," the speaker concludes, "What is it all but a trouble of ants in the gleam of a million million of suns?" (3, 4). Published a year later, "Locksley Hall Sixty Years After" has a similar ethos, though Tennyson claimed that the poem is a dramatic monologue. The poet looks back with disillusionment on a long life. He questions whether war will ever cease on the earth "till this outworn earth be dead as yon dead world the moon?" (174). And in the Epilogue to "The Charge of the Heavy Brigade at Balaclava" (1885), the "Poet" deflects the romantic notion that the stars are human souls (i.e., of fallen soldiers):

> The vast sun-clusters' gather'd blaze,
> World-isles in lonely skies,
> Whole heavens within themselves, amaze
> Our brief humanities. (53–56)

Tennyson's most memorable poems confirm the notion that nineteenth-century humanism was both progressive and tragic, an idea also readily observed in one of the Victorian era's most admired poems, Matthew Arnold's "Dover Beach," which the poet wrote around mid-century but was not published until 1867. A much-discussed poem with many interpretations, the final stanza, whether the thought of a moment or a final pronouncement, rejects the romantic sense of cosmic order that once prevailed because it is no longer believable. Humans are subject to the whims of uncontrollable outer forces that throw us around as randomly as waves fling the strand pebbles established in the first stanza. This tragic ethos also permeates Robert Browning's "Love among the Ruins" (1855), which, inspired by Shelley's "Ozymandias" and *Queen Mab*, is set in a pastoral scene, "the site once of a city great and gay" (7)—Browning's original manuscript is entitled "Sicilian Pastoral." Now, though, "the country does not even boast a tree" (14). The grass "o'erspreads/And embeds/Every vestige of the city" (27–29). Only "the single turret remains" (37); even here, nature is reclaiming the area: "the patching houseleek's head of blossom winks/Through the chinks" (41–42). The poem concludes, echoing the first lines of Arnold's final stanza ("Ah, love, let us be true/To one another!" (29–30)), that only love can help us transcend the weight of fleeting time:

> Oh heart! oh blood that freezes, blood that burns!
> Earth returns
> For whole centuries of folly, noise and sin!
> Shut them in,
> With their triumphs and their glories and the rest!
> Love is best. (79–84)

Among the bleakest of Victorian poets, James Thomson (not to be confused with the earlier writer of *The Seasons*) dramatizes London as a ruined city in *The City of Dreadful Night* (1874). The city has become a "Venice of the Black Sea," through which the River of Suicides flows. The narrator observes a couple speaking near the black river of the black city:

> Man might know one thing were his sight less dim;
> That it [the earth] whirls not to suit his petty whim,
> That it is quite indifferent to him. (9.42–44)

The poem revisits the idea in Part 13, as the narrator notes, echoing Epicureanism (and Spenser), that "naught is constant on the earth but change" (7). Though men are unwise in their use of time, "He naturally claimeth to inherit/The everlasting Future" (26–27). In his dark manner, Thomson covers territory (by way of couplets) established by Pope in *An Essay on Man*, though if there is any sense of order, it is mere chance. In fact, our lives are "little" (14.49), we die, never to wake (14.51), while life is an unending cycle, and we will give up our places "To other beings, with their own time":

> We bow down to the universal laws,
> Which never had for man a special clause
> Of cruelty or kindness, love or hate:
> If toads and vultures are obscene to sight,
> If tigers burn with beauty and with might,
> Is it by favour or by wrath of Fate? (14.61–66)

Thomson was not fond of Tennyson, but many of the ideas here share territory with *In Memorium*. The idea that "All substance lives and struggles evermore" holds implications for Darwinian science, though life is governed not by "natural selection" but "Necessity Supreme" (14.75). Also with ecological implications, the air is charged with human feeling and thought, and life is "overfraught . . . so that no man there breathes earth's simple breath" (15.7–8). Thomson's "poisoned air" (18) may work metaphorically, but it is also a pining for too strong a human footprint and a wish for simplicity. Nature is not, as it often is for Wordsworth and other writers, a lasting cure, an antidote to melancholy—"all is vanity and nothingness" (15.70) is almost the final word in the poem. One may wonder if Thomson's visit to Colorado during the time he composed the poem gave rise to his words in favor of nature, and the considerable air pollution in London at the time was certainly part of the poem's context.

Thomson's Swiftian proposal for "The Speedy Extinction of Evil and Misery" (1868–1871), which, like Hawthorne's 1844 story "The Earth's Holocaust" (see Chap. 5), caricatures reform proposals, lies in the extinction of man by mass suicide. Among the reasons for the proposal is our will to "poison our atmosphere" (24)—a central, recurring image in *The City of Dreadful Night*. Yet the Thomson of this essay is no pantheist or romantic lover of nature; the earth is imperfect, "poor" in its way (see 21, etc.) and uncaring, inhospitable to humans. It would be better if we all committed hari-kari (24). No one would confuse Thomson for Swift—or Twain, who was beginning to establish himself at this moment in the United States and would later show greater wit in making a case for the immolation of the human race.

As much as any writer of the late Victorian era, George Meredith was well-attuned to the science of his day, though he rarely mentions science directly in his work or in his letters. His nature poetry in particular marks a move away from the emotional and romantic toward the rational, as seen earlier in Arnold's "Dover Beach" and Tennyson's *In Memorium*. His poem "Meditation under Stars" (1888) is posited implicitly on a revocation of the anthropocentric viewpoint regarding the universe that is out of step with modern science. Addressing the "links" between humans and the distant, "implacable" planets and stars, no easy answers are available to us on Earth, "a shuddering prey/To that frigidity of brainless ray" (32–33). Science may seem to have made the distant stars isolated and isolating, "cold" (68), but this is an anthropocentric delusion. The poem suggests that we abandon the outdated view of the stars as sentient and, a holdover from antiquity and the Middle Ages, connected mystically to our lives as heavenly guides. As Tess Cosslett writes about the poem, "We must escape from the self-referring viewpoint, and we must achieve a spiritual insight into the hidden connection processes: then our relative smallness and insignificance in the Universe are no longer a source of despair, for we understand how even the smallest part belongs to the great whole" (117). The stars, though "may we read and little find them cold:/Let it but be the lord of Mind to guide/Our eyes" (59–61). Humans, "specks of dust upon a mound of mould" (65), are inextricably kindred to the stars—"Though low our place,/To them are lastingly allied" (66–67)—just as we are kindred to our own planet. Meredith concludes the poem by bringing the focus back to the earth. Unsentimental, nonanthropocentric reflection on our link to the vast universe may leave us with a beautiful wonder about our own domain: "Half strange seems Earth, and sweeter than her flowers" (83).

The French Symbolist Movement

Many of the ideas I have discussed apply to some of the French Symbolists, though in a different style and aesthetic sense. The Symbolists were largely *not* nature poets, preferring crowded Paris, with decadence on full display; like most twentieth century modernists, the Symbolists were city poets. For Baudelaire in particular, Wordsworthian rurality is not a source of poetic inspiration but a place to instill boredom. The Symbolists mark the earliest form of modernism, a chief characteristic of which is a strong sense of the loss of order and the loss of traditional values (combined with the loss of traditional forms and techniques). The denial of the primacy of humans, our central position in the universe, may also be seen as one aspect of this lost order, even though (as I have shown) this idea was very much in question since antiquity and even more so and in a wider sense with the discoveries of Copernicus, Galileo, and Newton, among other scientists. Contemporaneously with the middle and late Victorians, French writers reacted to discoveries in science, including Darwin, in similar ways, though, as Edmund Wilson argues, in a fashion that is (purposely) harder and less graceful than Tennyson (8). Early adherents to Parnassism, which espoused art for art's sake, such as Théophile Gautier and Gerard de Nerval held an open aversion to Saint-Simonian humanitarianism (Fowlie 2), and the chief philosopher for the Symbolists was Schopenhauer. The later French Symbolists also dismantle traditional humanism, though they are more interested in suggesting ideas through private, rather arbitrary imagery.

Nerval (1808–1855), after Charles Baudelaire the most important Symbolist precursor, rejected the sentimental romanticism of writers such as Victor Hugo.[18] For Nerval, as for, later Baudelaire and Stéphane Mallarmé, a work of art is built not on a subject matter but on absence: "The void left by a completed experience is the authentic subject matter of art, and in a literary work, words come to fill this void without, however, building a real substance" (Fowlie 17). In his dreamlike prose masterpiece "Sylvie," the unnamed narrator (evoking Rousseau, mentioned twice in the story) seeks escape from the city to the simpler country. Nearing a lake with a ruined temple near Ermenonville, he says, "Yes, this temple is crumbling like so many others, and man, weary or forgetful, will turn away from its threshold while nature, indifferent to all, reclaims the terrain that art tried to wrest from her; but the thirst for knowledge will live on for ever, the spurt of all vitality and all action!" (166). In what is his most famous poem, the sonnet "El Desdichado," from *The Chimeras*, Richard Lattimore translates its first two lines: "I am the dark, the widowed, the disconsolate,/I am the prince of Aquitaine whose tower is down." Wallace Fowlie gives an autobiographical reading of the poem: Nerval claimed he was in the lineage of the Labrunie knights, the leader of which was called "le duc d'Aquitaine." The family coat of arms consisted of three towers—in the poem concentrated into one, now fallen, indicating the poet's loss of a connection to his paternal lands (20–21).

Here and elsewhere the Symbolists indicate a continued interest, following Enlightenment and Romantic figures, in fallen (or falling) civilizations. Gautier writes that Baudelaire's decadent style "is nothing more than art arrived at that point of extreme maturity produced by aging civilizations when their sun begins to set" (qtd. in Calasso 283). One of Baudelaire's most anthologized poems, "Une charogne" ("A Carcass)," begins as a sort of love poem ("Remember, my love, the object we saw/That beautiful morning in June") and goes on to describe in opulent language the body of a rotting, stinking animal carcass. Near the conclusion, the love theme resumes in a blackly humorous turn: "—And you, in your turn, will be rotten as this . . . My passion, my angel in one!" (1–2, 37, 40). Fowlie suggests that the point of the poem is not the description of the putrefying carcass but the biblical idea that humans are made of and will return to dust (46). Echoing Baudelaire's images on decaying empires, another important French Symbolist, Paul Verlaine, addresses the falling of humanity from its teleological throne in "Langueur" ("Apathy"):

> I am the Empire at the end of its decadence
> Watching the tall, fair Barbarians pass,

Meanwhile, I compose idle acrostics
In a golden style where the sun's languors dance. (1–4)

Not only a protomodernist but also a harbinger of posthumanism, Verlaine was skeptical toward industrialism, which he believed would eventually destroy the human personality (Fowlie 108).

In the same decadent strain, but with a stronger sense of parody and self-ridicule, Jules Laforgue's (1860–1887) initial (but abandoned) first book of poetry, *Le Sanglot de la terre* (*The Tears of the Earth*), is strongly speculative and ambitious. His sonnet "Apotheosis" opens with an image of a vast universe where each star "twinkles in dismal isolation" down to a "yellow point, Paris," to the poet himself, "A weak phenomenon in the universal order." His bird's eye of history, "Funeral March for the Death of the Earth," apostrophizes an earth that (like Byron's "Darkness") the sun has blotted out and where "[t]ime has ceased": "You are now but a coffin, an inert and tragic block," but the emphasis is all on human history. Echoing ideas from Hesiod and Ovid, the Earth's "early ages" were comprised of "the spleen of long days," but then "impure being came, weak rebel,/And tore the lovely veils from holy Maia": Humans probed nature's secrets and defiled a nurturing mother nature. The Earth persisted through "the dark Middle Ages," during which, "to the disturbing knell of *Dies Irae*," famine and pestilence thrived, and man cringed, "obstinately begging for Mercy" and (presumably) through the eighteenth, "hysterical century, when doubting man/Found himself alone without Justice, without God,/Rolling in the unknown on an ephemeral globe." Much of the poem's second half is a commentary on the folly of human achievements, gods, histories, philosophy, literature, "Books, the record of man's futile victories." The dead Earth is now nothing but a tomb filled with "dark, time and silence." Meanwhile, the "solemn procession of magnificent suns"—the repeated chorus in the poem—continues unabated (pp. 11–13).

The great French realist Guy de Maupassant's story "The Horla" extends such thoughts in a tale mixing psychological horror, vampirism, and science fiction as the narrator, a mix of coldly rational, skeptical, and mad, relates in his diary the process by which he is overtaken by a demon that he comes to believe he has summoned unwittingly from a Brazilian ship sailing on the Seine. The story's theme of the unreliability of human senses and the utter weakness of the human body provides a pretext for the possession. Human eyes are relatively weak, and our other senses are unreliable; humans "are so weak, so defenseless, so ignorant, so small on this speck of dust spinning around in a drop of water!" (335). The invisible demon, the narrator concludes, is among the alien beings that will overtake humankind. Maupassant is not a didactic writer, but perhaps the story may be read as a cautionary against dismissing too forcefully our species as insignificant, the results of which may lead to a sort of madness. Our relative insignificance, the fact of our late invention (as Foucault puts it) does not determine despair or resignation. Much the same idea is reflected in Wordsworth's "Expostulation and Reply," in which Matthew chides his friend William for sitting too long outdoors dreaming, looking upon

> . . . mother earth,
> As if she for no purpose bore you;
> As if you were her first-born birth,
> And none had lived
> before you! (9–12)

To this, William replies,

> The eye it cannot chuse but see,
> We cannot bid the ear be still,
> Our bodies feel, where'er they be,
> Against, or with our will. (17–20)

The twentieth century, the bloodiest in history, would have much to say about the apocalypse, the end of history, and the relative smallness of humans. Scientifically, findings about the immensity of the universe (as it turns out, only hinted at in earlier centuries) gave logical evidence for if not outright proof of the groundlessness of anthropocentrism. Much of the melancholy of literature among realists and naturalists in the late nineteenth and early twentieth centuries revolves around the idea of the cosmic arrogance of humans. Anton Chekhov's story "The Lady with the Little Dog" (1899) is one example. At a sea resort in Oreanda (Ukraine), Gurov, a Moscow banker in early middle age, and Anna, half Gurov's age, rest between secret lovemaking sessions (their spouses are far away). They sit on a bench near the sea and observe the quiet scene:

> The leaves of the trees did not stir, cicadas called, and the monotonous, dull noise of the sea, coming from below, spoke of the peace, of the eternal sleep that awaits us. So it had sounded below when neither Yalta nor Oreanda were

> there, so it sounded now and would go on sounding with the same dull indifference when we are no longer here. And in this constancy, in this utter indifference to the life and death of each of us, there perhaps lies hidden the pledge of our eternal salvation, the unceasing movement of life on earth, of unceasing perfection. (366–67)

The somber, all-too-true but undidactic passage appears amidst a narrative set in a world of fashionable, bourgeois life, as Gurov ponders his own unhappy marriage and the realization that he is aging. He pursues and eventually wins an equally unsatisfied Anna, but Chekhov, the most committed of realists, leaves the couple's future uncertain.

Another late realist (or naturalist) sometimes grouped with modernists, Thomas Hardy's poem about the sinking of the *Titanic*, "The Convergence of the Twain" (1912), in another way, shows that the human presumption to controlling nature is folly and can even be deadly.

> And as the smart ship grew
> Its stature, grace, and hue,
> In shadowy distance grew the Iceberg too. (22–24)

As the liner and the iceberg converge, "consummation comes, and jars two hemispheres" (33). The final line in the above tercet may be seen as an example of what Ernest Callenbach would call the fourth law of ecology: "Nature bats last." A variant on this (altering Thomas à Kempis) might state, "Man proposes, nature disposes."[19]

Outside my present scope, I can only mention the growing sense of alienation, discussed by such social critics as Max Weber and Georg Simmel, in the early part of the twentieth century. The *fin de siècle* gave rise to the idea of the decline of the West. Henry Adams sensed a change in human perceptions while viewing the hall of dynamos at the Paris Exposition in 1900; he "began to feel the forty-foot dynamos as a moral force, much as the early Christians felt the Cross. The planet itself seemed less impressive, in its old-fashioned, deliberate, annual or daily revolution" (318). Early in the new century, Bertrand Russell wrote, "All the labours of the ages, all the devotion, all the inspiration, all the noonday brightness of human genius, are destined to extinction . . . The whole temple of Man's achievement must inevitably be buried beneath the debris of a universe in ruins" (37). The first world war brought about cultural crises and provided tangible signs of a civilization in decline with scores of wounded and scarred returning veterans and cities filled with ruins. Ezra Pound declares famously in *Hugh Selwyn Mauberley* that Western civilization is "an old bitch gone in the teeth" and "botched" (5.3, 4).[20] The French poet Paul Valéry, who grew out of the French Symbolist Movement, wrote in "The Crisis of the Mind" (1919), "We later civilizations . . . we too now know that we are mortal" (94). Through his organic (Goethean and Nietzschean) conception of history, Oswald Spengler, in *The Decline of the West* (1918–1923), prophesied the inevitable cultural collapse of Western civilization. *The Waste Land*, Eliot's watershed poem of a postwar West, sterilized and in decay, is a sort of fulfillment of the poetry of the French Symbolists, one of many sources in Eliot's interpenetration of fragments.

Notes

1. Lewis goes on to make a division between Christianity in which the redemption of man is central—history as the story of man's fall and redemption by a benevolent God—and that in which God is beloved and man is "a marginal creature" (120).
2. The classic study on Burnet and the rise of irregularity in aesthetics is Marjorie Hope Nicolson's 1959 book *Mountain Gloom and Mountain Glory: The Development of the Aesthetics of the Infinite* (Seattle: U of Washington P, 2011). For Burnet's influence, see especially Chaps. 5 and 6.
3. For more on the philosophical-literary theme of "nature loves to hide," see Pierre Hadot's *The Veil of Isis: An Essay on the History of the Idea of Nature*. Trans. Michael Chase. Cambridge: Belknap, 2008.
4. Discussing Spenser's celebration of order within what seems like constant change, John Watkins writes that Pythagoras' sermon at the end of Ovid's *Metamorphoses*, from which Spenser draws, "voices Ovid's belief in change as the paradoxical constant of cosmic existence" (97).
5. Regarding the excessive hyperbole of "Anatomy," Ben Jonson remarked that Donne's praise would have been appropriate if the subject had been the Virgin Mary, but not so a fifteen-year-old girl. Of course, Elizabeth Drury serves as a trope for decay that the poet (and all of us) lament. In the poem, writes Achsah Guibbory, Donne "describes a world which has lost integrity, health, wholeness, spirit, godliness, virtue. It was as if he and the few who had any memory of virtue left were lonely survivors, anticipating a cosmic, catastrophic end" ("John Donne." *The Cambridge Companion to English Poets*. Ed. Claude Rawson. Cambridge: Cambridge UP, 2011: 111).
6. For more on the topic, see Bridget Keegan, *British Labouring-Class Nature Poetry, 1730–1837*. New York: Palgrave Macmillan, 2008.

7. The poetic devotion to Newton was not mutual. Writes Bush: "The poets . . . thought far more highly of Newton than Newton thought of poetry" (55).
8. In the notes to his 1755 edition of *An Essay on Man*, William Warburton writes that in the poem Pope seeks to "reprove" the interpretation of the Bible placing man as a tyrant over animals. Man, Warburton writes, "soon began to consider the whole animal creation as his slaves rather than his subjects; as being created for no use of their own, but for this only: and not so content, to add insult to his cruelty, he endeavoured to philosophize himself into an opinion that animals were mere machines, insensible of pain or pleasure" (Alexander Pope. *An Essay on Man . . . with the Notes of Mr. Warburton*. London, 1755: 69.) Robert P. Irvine suggests that this idea may have had an effect on Robert Burns in his composition of "To a Mouse" (Burns 309 Note 72).
9. One of the earliest and most often cited works that problematize the neat binary of the rational eighteenth century poetry versus the humanitarian, nature-oriented poetry of Wordsworth and Coleridge is Robert Mayo's "The Contemporaneity of the Lyrical Ballads," *PMLA* 69 (1954): 486–522. Paul de Man deconstructs this binary in his influential essay "The Rhetoric of Temporality," *Blindness and Insight*. 2nd ed. Minneapolis: U of Minnesota P, 1983. 187–228.
10. Space prevents me from comparing Clare's poem with many of Seamus Heaney's poems, including "Bogland" and "The Tollund Man," which also evoke place and geological time powerfully by excavating the past. In the former poem, every layer "Our pioneers keep striking/Inwards and downwards,/Every layer they strip/Seems camped on before" (*Selected Poems 1966-1987*. New York: Noonday, 1990. ll. 23–26). Heaney discusses Clare in "John Clare, Prog" (*The Redress of Poetry*, New York: Farrar, 1996: 63–82.)
11. In his *Alpine Journal* (written the same year as *CHP* III), Byron clearly shows himself as a lover of wild nature, an appreciator of the sublime—a glacier, he notes, resembles a "frozen hurricane" (987). Being in the wilds enforces a hatred of civilization; on the road from Thoun to Bern are "good road—hedges— . . . prosperity—and all sorts of insipid civilization" (988). Reflecting *CHP* as well as *Manfred* and *Cain*, he writes, "I am a lover of Nature," but this has not "lightened the weight upon my heart—nor enabled me to lose my own wretched identity" (990). As an animal lover, Byron owned, for better or worse, a pet bear, as well as a macaw and parrot; writes Marchand, he kept a menagerie of animals around him to escape exigencies of human relationships (46, 166).

12. McCann writes, "I cannot agree with [Gleckner's] controlling idea that Byron's is a poetry of radical despair. His poetry is indeed built upon a vision of the world's horror and absurdity, and while he never for a moment forgets this vision, his later poetry frequently advances alternative insights and possibilities" (ix).
13. In his travelogue "Ktaadn," Thoreau writes of ascending Mt. Ktaadn, on the summit of which he finds himself in the hostile territory of a blinding, windy mist. He imagines Nature, here a "stepmother," speaking to him: "Shouldst thou freeze or starve, or shudder thy life away, here is no shrine, nor altar, nor any access to my ear" (*The Maine Woods.* New York: Penguin, 1988: 86).
14. Cf. Cormac McCarthy's *The Road* 32.
15. According to scientific estimates, the sun will not burn itself out for around five billion years. See Jill Scudder. *The Conversation.* "The sun won't die for 5 billion years." 13 Feb. 2015. http://phys.org. Accessed 9 Dec. 2016.
16. See Frank Kermode and John Hollander, eds. *The Oxford Anthology of English Lit Vol 2.* New York: Oxford UP, 1973: footnote, 308. In a review of Campbell's "The Last Man," Francis Jeffrey stated that Campbell's poem was indebted to Byron's "Darkness." This prompted Campbell to write an open letter claiming that it was *he* who had suggested the subject to Byron at least fifteen years before "Mary Shelley's The Last Man: Apocalypse without Millennium." *Keats-Shelley Review* 4 (1989): 1–25.
17. Mab is the fairy subject of an elaborate but lighthearted speech by Mercutio in *Romeo and Juliet* 1.4.
18. Yet even Hugo suggests a decentering of humanity in his untitled poem "Do you take mankind for be-all and end-all?," which is aimed at Urbaine Leverrier, discoverer of the planet Neptune and loyal supporter of Napoleon III, whose rise to power caused Hugo to voluntarily seek exile. See *Selected Poems of Victor Hugo: A Bilingual Edition.* Trans. E.H. and A. M. Blackmore. Chicago: U of Chicago P, 2001. 553, 612.
19. Callenbach's laws are almost identical to ecologist Barry Commoner's four laws of ecology in *The Closing Circle: Nature, Man, and Technology* (1971): (1) Everything is connected to everything else; (2) everything must go somewhere; (3) nature knows best, and (4) there is no such thing as a free lunch. See Michael Egan, *Barry Commoner and the Science of Survival: The Remaking of American Environmentalism.* Cambridge. MIT, 2007: 126–127). Fifteenth-century theologian Thomas à Kempis uses the Latin phrase "Homo proponit, sed Deus disponit" ("Man proposes, God disposes") in Book I of *The Imitation of Christ.*

20. Pound attacked anthropocentrism as early as 1910: "Man is concerned with man and forgets the whole and the flowing" (*Spirit of Romance* New York: New Directions, 1968. 93). Feng Lan writes that Pound suggests the Confucian "jing"—i.e., roughly, respect for nature—as its antidote in No. 88 of his *Cantos.* (*Ezra Pound and Confucianism: Remaking Humanism in the Face of Modernity.* Toronto: U of Toronto P, 2004: 158).

CHAPTER 4

Teleology, Ecology, and Unity and the French Enlightenment

The Enlightenment has fallen on hard times in the twentieth and early twenty-first centuries. In 1932, two roads diverged in a modern reassessment of the movement's status. Ernst Cassirer (in *The Philosophy of the Enlightenment*) demonstrated that the philosophes created modern philosophy, while Carl L. Becker (in *The Heavenly City of the Eighteenth-Century Philosophers*) eviscerated the philosophes' overreliance on reason and logic, the latter being "something the mind has created to conceal its timidity and keep up its courage" (25). Becker argues that the philosophes are closer to the thought of the Middle Ages than that of modernity. In *Dialectic of Enlightenment*, Max Horkheimer and Theodor Adorno write of the "disenchantment" of the Enlightenment based on what Zuidervaart Lambert summarizes as a "an irrational fear of the unknown" that seeks for "blind domination . . . of nature by human beings, the domination of nature within human beings, and, in both of these forms of domination, the domination of some human beings by others."[1] Echoing ideas by Horkheimer and Adorno, others assert that the Nazi death camps are the result of such blind domination.[2] For many poststructuralists, when the philosophes, for whom reason is the metanarrative, are discussed at all, they are characterized as essentializers concerned with intentionality and unaware of the temporality and rhetoricity of language. Some feminist and postcolonial scholars attack the movement as enforcing and spearheading homogeneity at the cost of the degradation of women and people of color.

Yet other twentieth and twenty-first century writers have vigorously defended the Enlightenment and counter that it is far less out of step with

B.L. Moore, *Ecological Literature and the Critique of Anthropocentrism*, DOI 10.1007/978-3-319-60738-2_4

contemporary theory and more politically aware than poststructuralism in general. Jonathan Israel calls Nazism "the supreme Counter-Enlightenment" (xi). Jean Starobinski takes to task the notion that a writer like Voltaire held an essentializing viewpoint in his *contes*: "The Voltairean tale offers us an accelerated, caricatural image of this constant motion, of this oscillation between nature and culture, vice and virtue, laughter and tears, pessimism and optimism, and leaves us in a state of overall confusion despite all the clarity of the individual details" (117). Peter Gay, one of the most important twentieth-century Enlightenment apologists, suggests that most of the Enlightenment hostility toward Judaism is rooted in an attempt to enlarge a "world once circumscribed by faith" (*Enlightenment: The Science* 391). Karen Offen writes that "the European Enlightenment is far richer in content and scope on gender issues, indeed far more explicitly 'feminist' in its claims and aspirations than has been generally acknowledged" (99).[3]

Attacked for its colonialist views, some of the key Enlightenment writers were skeptical and openly critical of European claims of superiority over native peoples. In *Candide*, for example, the title character meets a black slave whose master cut off a hand for having caught a finger in the sugar mill and a leg and trying to escape. Says the slave to Candide, "It is at this price that you eat sugar in Europe" (60). Such a statement—one of the few that causes the naïve title character to renounce (at least temporarily) the absurd reasoning of his "oracle" Pangloss—holds strong implications in new ways today for global capitalism. Echoing Montaigne's "Of Cannibals" and anticipating Melville's *Typee,* Diderot's *Supplement to Bougainville's "Voyage"* argues forcefully against slavery, colonialism, and the standard notion that Western cultures are superior to non-Western ones.

This chapter seeks less to credit or discredit the attackers or defenders of the Enlightenment than to show how a handful of its more celebrated literary works come into play with regard to the notion of a human-centered universe. In this regard, the philosophes' views about the credibility of a human-centered world seem to me to be quite progressive, scientifically aware, "enlightened," and with important implications for posthumanism. The romantics rejected uniformly what they saw as a mechanical, Newtonian view of the universe as established by the French philosophes, and the failures of the French Revolution caused the romantics in general to turn away from the political deeper into mental experience (see Breckman 16), but romanticism was also a reaction to the displacement of the human subject so typical of Enlightenment writing. Yet this turn—away from teleology and anthropocentrism—is actually central in a number of French Enlightenment works. I focus here on two very early figures of the movement, Fontenelle and the English writer Shaftesbury, and then proceed to a discussion of Voltaire, Rousseau, and Diderot. Though the idea, especially in its development into outright ecocentrism, would receive vitally important development in the work of later scientists and writers, the Enlightenment modernized the response to anthropocentrism by dissembling its bases through scientific reasoning and wit.

Kant wrote that "Enlightenment is man's emergence from his self-imposed nonage"—the final word is sometimes translated "immaturity" or "minority"; this is, he continues, "the inability to use one's own understanding without another's guidance" ("What Is Enlightenment?" 1071). The Enlightenment also, in a sense, eroded the centrality of humankind in the universe, away from the mystical to the scientific. It freed man from his unwillingness to think for himself and elevated human reason in ways unrealized before, but it also identified man as not the measure. Commenting on the "shift in the locus of essential secrets from God to nature," Evelyn Fox Keller writes, "Over time, the metaphorical import of this shift was momentous; above all, it came to signal a granting of permission to enquiring minds—permission that was a psychologically necessary precursor for the coming Enlightenment. Indeed, Kant's own answer to the question 'What is Enlightenment?' was simply this: 'Sapere aude'—dare to know" (98).

Conversations on the Plurality of Worlds by Bernard de Fontenelle (1657–1757) is precisely such a daring to know, though the book appeared in 1684, forty years before Kant's birth. A demystifying work, its influence marks the early phase of the Enlightenment. Fontenelle wrote the book as a piece of women's entertainment and is comparable to popular science books of the twentieth and twenty-first centuries. The book does not go into a lot of detail about astronomical science, but in context it shows a growing skepticism toward received knowledge, as does Fontenelle's *Philosophical Dialogues of the Dead*, based on similar dialogues by Lucian. *Conversations* appeared during a fertile period in science literature, during the same decade as Newton's first edition of the *Principia* (1687), which revolutionized physics and many other disciplines.[4]

The world of *Conversations* is pre-Newtonian as well as pre-Herschelian, though the book does, in its way, communicate the idea of a vast universe. The controlling worldview is, in fact, Descartes' mechanical world of Vortices; one of the controlling metaphors is that of a clock or a watch (e.g., 21). Many of the ancients, including Pythagoras, suggest or state explicitly the possibility of a plurality of worlds, though this idea is for Fontenelle mostly secondary to a discussion of astronomic principles in general. Fontenelle attacks the idea of an anthropocentric universe near the start of the book, and he continues to do so repeatedly and in myriad ways for the rest of the way. Over six evenings, Fontenelle teaches and converses with a beautiful woman, a Marchioness identified as Madame de la Mesangire of Rouen. Part of the work's fun is Fontenelle's treatment of serious subject matter in a light, slightly flirtatious context that merges with courtly backdrop of the day. So, he says, our point of view as humans is severely limited, "as spectators of an opera" (20).

Carolyn Merchant notes the "condescending" tone of the male teacher toward the Marchioness (272). The book certainly does little to subvert male hegemony, yet its tone is more playful and comic at the expense of educating a flirty court woman.[5] Far from foolish, the Marchioness echoes Donne's "First Anniversary," in which "The sun is lost, and th' earth, and no man's wit/Can well direct him where to look for it" (207–08). For the poem's speaker, as for the Marchioness, as for those operating with a bad education, or in ignorance, complacency, or the lack of educational opportunities, understanding of the world may lead to a certain vertigo, an unsteadiness, and confusion in the realization that, at least temporarily, as Donne put it earlier in the century, "all coherence is gone" (213).

The narrator writes that "we chuse to believe that every thing in creation is destined to our service." Why are there so many stars in the night sky? So that our sight might be gratified? This is a "selfish principle," he writes, that is linked to our ignorance. Whereas formerly the earth alone was thought to be "motionless in the midst of the universe," the "sole purpose" of other heavenly bodies was to light up the earth (24). For her part, the Marchioness prefers a Ptolemaic universe because it is more comforting, less complicated. But the narrator replies that the same disposition that makes "a man of the world to aspire after the most honorable place in the room, will make a philosophy desirous of placing the globe on which he lives in the most distinguished situation in the universe." To the Marchioness' charge that he is "calumniating human nature," the narrator replies regarding Copernicus' theories that humiliating is not same thing as making one humble (29). Copernicus' theories, she counters, do not make her feel humble at all (28); she feels that she is the center of the universe and is not in motion (33). Replying to the narrator's statement that "[t]he motions of self-love are so frequent in our minds, that for the most part we are not sensible of them," the Marchioness tells him to quit moralizing (35). Later she tells the narrator that he is "making the universe so unbounded that I feel lost in it; I don't know where I am, nor what I'm about . . ." She finds it difficult to comprehend the idea that our sun and planets are "but a little portion of the universe!" The idea of other, similar worlds is "fearful; overwhelming!" The narrator, by contrast, finds it pleasing; creation is "boundless in treasures; lavish in endowments" (94).

Anticipating a number of later science fiction stories (and echoing a few earlier stories, as from Lucian), the narrator speculates that the inhabitants on some planets in the Milky Way feel sorry for earthlings, "miserable creatures who spend half their time in profound darkness" and who have "but one sun." The inhabitants of other planets "would think we had fallen under the displeasure of nature" (98). Not to be completely negative about humans, the narrator states that human extinction will never occur, and the dialogue ends with an affirmation of the genius and unequalled taste of current Europe (105, 118–19). Near the end of the work, in an attempt to help readers consider a nonhuman point of view, the narrator suggests that comets may be inhabited by sentient beings (102).

Fontenelle's discussion of comets is particularly relevant. Observation of their movements was central for Newton and others in showing the mechanical operations of the cosmos and establishing some of the laws of physics (see Jardine 11–41). Fontenelle was, with Pierre Bayle, the great skeptic in the late seventeenth century regarding comets as portents of human affairs. Fontenelle's 1681 one-act comedy *The Comet* (*La Comète*) targets these long-held superstitions. A young man, M. de la Forest, wishes to marry Florice, the daughter of a wealthy astrologer, who overrules the marriage plans because of the Great Comet of 1680—a comet that would play a large role in the history of modern astronomy (and a portent in the decline of astrology). The father forbids the marriage because the would-be groom (more or less representative of Fontenelle) has such "outrageous" views that comets are not tied to human action but are natural events. The farce shows that it is the astrologer (and all others who view comets with superstition) who is the "outrageous" one.[6]

As one of the earliest champions of the new modern climate of the late seventeenth century, Anthony Ashley Cooper, 3rd Earl of Shaftesbury (1671–1713) was a considerable influence on French Enlightenment figures, especially Diderot, who translated (or paraphrased) his works into French as a young man. For Shaftesbury, as Pope puts it, the revelations of science, though they show man's "time a moment, and a point his space" (*Essay on Man* 72), are not a cause for dark thoughts: "Science, which may nourish man's pride, may also, with religion, nourish proper humility" (Bush 62). Reacting against the scientific philosophy of his own tutor, John Locke, and the Royal Society as well as the materialism of Hobbes, Shaftesbury did not approach nature scientifically but aesthetically and intuitively. Though his views are heavily deistic, for Shaftesbury, God is involved in his creation. His belief in the inherent goodness of man anticipates in short term the (rather nationalistic) love of nature in James Thomson and Thomas Warton, and it influenced directly not only Diderot but Rousseau, Goethe, Kant, Herder, and Montesquieu, to name only some of the more famous neoclassical figures. It fell out of favor for Locke and his successors but reemerged in the romantic era.

In his major work, *Characteristics of Men, Manners, Opinions, Times*, Shaftesbury's hero Theocles says, "All things in this world are united. For as the branch is united with the tree, so is the tree as immediately with the earth, air, and water which feed it . . . Thus in contemplating all on earth, we must of necessity view all in one, as holding to one common stock. Thus too in the system of the bigger world. See there the mutual dependence of things!" Anticipating instinctively the concept of symbiosis in ecology, he writes that it is only by surveying the world that one may come to know "a union thus evidently demonstrable, by such numerous and powerful instances of mutual correspondency and relation, from the minutest ranks and orders of beings to the remotest spheres." The fact that the "mighty union" of things in nature are not easily seen is our source of wonder: "since each particular has relation to all in general, it can know no perfect or true relation of any thing in a world not perfectly and fully known" (2.64–65). The evidence of a "uniform consistent fabric" points to the existence of a "universal mind" that we cannot disown, "except through the imagination of disorder in the universe, its seat" (66).

His sunny view of creation includes man, though we are only a small part of creation:

> All is delightful, amiable, rejoicing, except with relation to man only, and his circumstances, which seem unequal. Here the calamity and ill arises, and hence the ruin of this goodly frame. All perishes on this account; and the whole order of the universe, elsewhere so firm and entire, and immovable, is here overthrown and lost by this one view, in which we refer all things to ourselves, submitting the interest of the whole to the good and interest of so small a part . . . (67)

Less sunny than Shaftesbury, Voltaire's *Micromégas: A Philosophical Story*, inspired by *Gulliver's Travels*, shows a picture of the eroding centrality of humankind in the universe in the wake of Newton. Published in 1752, the essential elements of Voltaire's *conte* were contained in the account, now lost, of an imaginary voyage sent to Voltaire's patron and (then-) friend Frederick of Prussia in 1739. Micromégas, a 120,000-foot-tall inhabitant of a planet orbiting Sirius, visits Saturn, where the inhabitants stand at a mere one thousand fathoms (1800 m). Micromégas laments the fact that Sirians possess only one hundred senses, "for we in our globe have nearly a thousand senses, and still there remains in us I know not what vague desire, what uneasiness, that incessantly reminds us that we are nothing much and that there are beings much more perfect" (176). By implication, humans are in a relatively very lowly position, limited to five senses, standing an average of a bit over five feet tall, and living only seventy years.

Visiting earth, Micromégas, accompanied by a Saturnian "dwarf," finds a tiny creature in the Baltic Sea, a whale. Encountering a boat filled with atoms, in fact humans, who it turns out are philosophers, apparently talking to one another, Micromégas fashions a speaking trumpet from his fingernail, converses with them, and discovers that the "insects" are rational. Among the philosophers, an absurdly tiny human adherent of the Catholic philosophy of Thomas Aquinas prompts ridicule by asserting the (hard) anthropocentric notion to Micromégas and the Saturnian "that their persons, their worlds, their suns, their stars, everything was made solely for man" (190). Micromégas and the Saturnian can only respond to such a ridiculous notion with convulsive laughter. In the Newtonian universe, it is the law of nature to (as the Saturnian says), "feel like a drop of water in an immense ocean" (177).

Voltaire attacks anthropocentrism in many other works, including his "Dialogues between Lucretius and Posidonius, First Colloquy." Posidonius, a Stoic and believer in final causes, says that the argument for a designed world by an intelligent supreme being whose ends are to benefit man has many obvious proofs in the order of nature. Lucretius, an Epicurean who denies rational design, responds with the proto-evolutionary notion that, over time, matter can produce an *Iliad* (77). Nature, he says, modifies itself and "requires millions of ages . . . to arrive at last at the only one which can produce living beings" (78). He attacks Posidonius' statement that ours is "the best of all possible worlds," an idea central in Voltaire's most famous work, *Candide*, in which Pangloss says, "Stones were formed to be cut and to make into castles; so my Lord has a very handsome castle" (16). A central event in the writing of *Candide*, and a part of its plot beginning in Chap. 5, is the 1755 earthquake that destroyed Lisbon and killed more than 30,000 of its people. The event caused Voltaire and others to ask questions about a divine design of nature and the place of man in it: how could such a terrible event be a sign of divine order or of a benevolent God? The quake prompted, for many, the pressing need to "overcome" nature. The quake also prompted Voltaire's *Poème sur le désastre de Lisbonne* (*The Lisbon Earthquake*), which, like *Candide*, savagely burlesques the optimism of Leibnitz and Pope: "we are not more precious in the eyes of God, than the animals by whom we are devoured," he writes in the poem's preface (557).

As in *Micromégas*, Voltaire attacks anthropocentrism by way of hyperbole in his brief *conte* "Plato's Dream," in which "the great Demiurge, the eternal Geometrician" sends genii out to create their own worlds. Demogorgon, an evil divinity, creates the Earth and, expecting praise from his colleagues, is proud that he has with his bit of mud created a "masterpiece," but the other genii respond only with "hoots." A genie fond of joking (echoing Hume) critiques the planet's many imperfections, as well as its curious "two-footed animal," which has "so many enemies and so little defense, so many maladies and so few remedies, so many passions and so little wisdom" (226). The other genii are mocked similarly for the imperfections marking other planets in the Solar System. The Demiurge concludes that the genii have created some good and some bad and that their creations "will last only a few hundreds of millions of years, after which, having learned more, you will do better" (227). Voltaire attacks anthropocentrism more directly in the twenty-fifth of his *Philosophical Letters*, which critiques Blaise Pascal's belief, stated in his *Pensées*, that God gave man an exceptional place in creation. For Voltaire, such an idea is too great a conceit; man is merely a part of nature. Pascal, writes Voltaire, is "the sublime misanthropist" in assigning all sorts of evils to humans (119). Voltaire agrees with him that human nature is mixed with good and bad. But life is worth living if one knows how to live well, outside of a metaphysical system.

Though not as adamant in his dismissal of final causes, Jean-Jacques Rousseau (1712–1778) suggests a more vital love for wild nature and veers closer to ecocentrism in his last book, *Reveries of the Solitary Walker*, than either Voltaire or Diderot. Much of the book is devoted to coming to terms with real and perceived attacks on him by his enemies (Diderot, Grimm, and others). By the Fifth Walk (out of ten), he turns to the idea, at least temporarily, that there is inherent worth in all things, that everything has its own niche (51). The Seventh Walk is key in expressing his love for botany and solitude in relatively wild places.[7] Botany was beginning in the eighteenth century to be pursued for own sake, independent of pharmacology (Hankins 114). What begins as a fairly conventional picturesque passage develops into a poetic expression—anticipating Goethe and Emerson—of the unity of nature, a system that includes humans:

> The more sensitive the observer's soul, the more he delights in the ecstasy aroused in him by this harmony. On such occasions, a sweet and deep reverie takes hold of his senses, and he loses himself with delicious intoxication in the immensity of this beautiful system with which he feels at one. Then all individual things escape him; everything he sees and feels is in the whole. Some particular circumstances have to restrict his ideas and limit his imagination for him to be able to observe the separate parts of this universe which he was striving to embrace in its entirety. (71)

A focus of Wordsworth a few decades later in much of his poetry, for Rousseau, reveries in the wilds renew us in ways not possible in the city. Rousseau's words are as close to those of a romantic as any written before his time: "reverie revives and amuses me, thought tires and saddens me . . . my soul roams and takes flight through the universe on the wings of the imagination in ecstasies that exceed all other pleasures" (70).

Rousseau goes further in asserting an antiteleological, (proto)biocentric defense for the existence of natural things for themselves in his opposition to a utilitarian, anthropocentric use for plants. Citing Theophrastus and Dioscorides, he notes that some people have "the habit of seeing plants only as a source of drugs and medicine." But, he continues, "Nobody imagines that the structure of plants could deserve some attention in its own right" (72). A contemporary reader does not have to look far—perhaps merely to Capitol Hill—to find powerful people who appear believe that Alaska exists mostly *for* its petroleum or that the Appalachian Mountains exist *for* coal, to cite only two extreme but very real examples of anthropocentrism in action. Views such as these, he writes, "which always relate everything to our material interest, which make us seek usefulness or remedies everywhere, and which would make us look at the whole of nature with indifference . . . are ones I have never shared" (74). When we look at flowers and other objects in nature solely for their human use, our attitude toward nature cannot help but change. This extends to the miner, who, unsatisfied with the sunshine and set on finding riches "buries himself alive, and rightly so, since he no longer deserves to live in the light of day" (76). In his love of walking in solitude and in his near-misanthropy, Rousseau reads in these and other passages rather like an early Thoreau, who also wrote of wilds while largely sauntering in woods not far from civilization. Of course, Rousseau is cited regularly as the most important romantic precursor, though the claim is by no means as settled as some suggest.[8]

Another writer who is sometimes associated with romanticism (and was a friend then enemy of Rousseau) is Denis Diderot (1713–1784). Although the romantics would reject what they saw in the philosophes as an overly mechanical—if not "dead"—view of nature, Diderot argues consistently that nature is not a machine but organic, forming continually, so much so that he is unusually skeptical about man as the end of existence. His greatest statement about what would be in the nineteenth century the field of biology, as well as many other future sciences, is contained in his work *d'Alembert's Dream*. The characters of the work, in dialogue form, are real persons, and Diderot's own views are installed at various points in all of them. Jean le Rond d'Alembert falls asleep early in the work and mutters ideas aloud while Mlle. de L'Espinasse (Julie) attempts to record them. The celebrated physician Theophile de Bordeu enters the room later in the work, and he tries to interpret d'Alembert's words for Julie. Unpublished until 1830 (almost fifty years after the author's death), it is one of Diderot's boldest works. A twenty-first century reader may be surprised not only at its wide range of topics but also its depth and speculative power. Its spirit is comic, yet topics range from what would later be called genetics (135), biology, including evolution theory, psychology (160), ecology (including Lovelock's Gaia theory), philosophy (it attacks Berkeley directly, modifies/borrows from and then counters Descartes—there is no mind, only matter, says Diderot), and it rejects formal logic over observation (105). Over the course of the work, Diderot breaks down of the differences between humans and other animals, their differences being merely a matter of organization. Diderot's response to anthropocentrism is implicit and is based wholly on biology. His understanding of the unity of nature is spread throughout the entire work, as when he (through Dr. Bordeu) posits that one may see in a single drop of water in a microscope the history of the entire world. D'Alembert says (while in a fevered sleep), "Who knows how many races of animals have preceded us? Who knows how many will follow the races that now exist? Everything changes, everything passes away—only the Whole endures" (117).

The question of how inorganic matter can become organic is central to *d'Alembert's Dream*. Diderot's pan-vitalist answer is that all matter is sentient. He establishes his central argument early in the piece, in reply to d'Alembert's statement that one "can't make something out of nothing." Diderot says, referring to the process by which d'Alembert himself—an illegitimate child who would someday become a widely-respected mathematician—came into existence: "the first rudimentary beginnings of my mathematician were dispersed throughout the delicate young bodies of his future parents," then the seed of d'Alembert "grows and develops by stages into a foetus. At last the moment arrives when it is to leave its dark prison. Behold the newborn child, abandoned on the steps of the church of St.-Jean-le-Rond from which he will take his baptismal name . . ." The child grows, "the result of eating and of other purely mechanical operations" (96). Through Bordeu, Diderot will propose the proto-evolutionary idea that an organism may evolve according to needs, just as the efforts of generations of armless people could result in an eventual growth in arms in succeeding generations. Creatures develop "in response to necessity and habitual use" (123). Diderot's argument in *d'Alembert's Dream* develops ideas from an earlier work, *Thoughts on the Interpretation of Nature*, in which Diderot asks, "Just as in the animal and vegetable kingdoms, an individual comes into being, so to speak, grows, remains in being, declines and passes on, will it not be the same for entire species?" Over time, species acquire "movement feeling, ideas, thought, reflection, consciousness, feelings, emotions, signs, gestures, sounds, articulate sounds, language, laws, arts and sciences" and so on (75).

Though mechanical, Diderot's view of the universe, writes P.N. Furbank, is constantly "in flux, producing ever-new combinations and transformations and in which all separate identities are dissolved or prove illusory" (337). Humans are a part of nature and are not in a privileged position on the natural scene. As Bordeu states, "Nothing that exists can be either against nature or outside of nature" (*d'Alembert's* 172). Although Voltaire, Rousseau, and Diderot, among other philosophes, opened up a can of worms by interrogating the idea that humans are in an elevated position in nature, they understood the unique ability of humans to reason, to ask the right questions, and to solve problems. In much the same spirit as the eighteenth-century philosophes, Loren Eiseley, another accomplished practitioner of and writer on science, celebrates the mysteries of nature, but he also celebrates the human mind's ability to better understand our place in the natural world: "the mind in the course of three centuries has been capable of drawing into its strange, nonspatial interior that world of infinite distance and multitudinous dimensions" (93).

Is Diderot, then, among the antihumanists? In one respect, he is not. He is, with other Enlightenment thinkers, preeminent in making humans the central focus of the world. Without man, he writes in an *Encyclopedia* entry, nature "would be nothing more than a scene of desolation and silence" ("Articles" 25). But his opposition to Cartesian mind-body dualism, echoing Spinoza's monistic universe, resembles the posthumanist basis for rejecting anthropocentrism: humans are objects along with everything else in the universe. Here again is an example of a trenchant mind capable of seeing and realizing ideas outside of the controlling narrative—the primacy human reason. Much the same operation applies to the great humanist Montaigne, whose motto was "What do I know?" and by the great English Enlightenment figure Swift, whose *Gulliver's Travels* appears to all but destroy human reason as mere folly.

This chapter could easily have included many other French thinkers of the period, central to and on the fringes of the Enlightenment. For example, Marquis de Sade, in the latter category, writes in *Juliette*, addressing (as was his inclination) the sexual act: "Nature stands in not the slightest need of propagation; and the total disappearance of mankind . . . would grieve her very little, she would no more pause in her career than if the whole species of rabbits or chickens were suddenly to be wiped off the face of the earth" (67). Critical reaction to de Sade has been varied, including attackers (Horkheimer, Adorno, and Andrea Dworkin) and defenders (Susan Sontag and Angela Carter). He claimed to be an heir of the Enlightenment, but, according to Peter Gay, "he was never more than a caricature" of the movement (*Enlightenment: The Rise* 25). Whatever his status, he is certainly among the French thinkers of the era who subvert the idea of final causes in nature. In any case, those who label the French Enlightenment thinkers as absolutely oblivious to antihumanist ideas do so in error.

Notes

1. According to Zuidervaart, "Contrary to some interpretations, Horkheimer and Adorno do not reject the eighteenth-century Enlightenment."
2. For example, George Bernard Shaw stated that Nazism is the "legitimate heir to the European Enlightenment" (qtd. in Gray 94). See also Arthur Hertzberg, *French Enlightenment and the Jews: Origins of Modern Anti-Semitism* (New York: Schocken, 1971: 10) and Docherty (in Gordon 216).
3. Among more recent titles generally affirming the Enlightenment, see James MacGregor Burns, *Fire and Light: How the Enlightenment Transformed Our World*. New York: St. Martin's 2013, and Anthony

Pagden, *The Enlightenment: And Why It Still Matters*. New York: Random, 2013. I do not suggest that the Enlightenment is only a French movement. Roy Porter, among others, shows that England has an equal intellectual claim. See his *Enlightenment: Britain and the Creation of the Modern World*. London: Penguin, 2000.

4. French astronomer Jérôme de Lalande, who knew Fontenelle, wrote a critical account and notes for Fontenelle's book, and Elizabeth Gunning translated the book into English in 1803. Earlier English translators of the work include John Glanvill (1687) and Aphra Behn (1688).
5. In his *Philosophical Dialogues of the Dead*, many women speakers, including Sappho, Dido, Fulvia, Margaret of Austria, and Elizabeth I, more than hold their own with their male interlocutors.
6. See James Howard Robinson, *The Great Comet of 1680: A Study in the History of Rationalism*. Northfield, 1916. The monograph is old but still helpful. Pierre Bayle's most important contribution to the knowledge of (and skepticism about) comets is *Various Thoughts on the Occasion of a Comet*. A modern translation is Robert C. Bartlett, Albany: SUNY P, 2000.
7. Letters Rousseau wrote during the early 1770s on botany were collected posthumously and published as *La Boutique* in 1800 and 1802 and, a few years later, in an edition with illustrations by P.J. Redouté. An English translation is *Pure Curiosity: Botanical Letters*. Trans. Kate Ottevanger. New York: Paddington, 1979.
8. Maurice Cranston claims Rousseau "is the first of the Romantics," and many others agree, including Walter Raleigh, who calls him "the father of the literary Romantics," and Irving Babbitt (see Cranston's *The Romantic Movement*. Oxford: Blackwell, 1994: 1 and Seamus Perry, "Romanticism: The Brief History of a Concept." *A Companion to Romanticism*. Ed. Duncan Wu. Malden: Blackwell, 1998: 7). Stephen Bann writes that Rousseau's influence on English and German Romanticism is "crucial" ("Romanticism in France." *Romanticism in National Context*. Ed. Roy Porter and Mikuláš Teich. Cambridge: Cambridge UP, 1988: 244). On the other hand, Isaiah Berlin calls him "one of its fathers" but tends to downplay him in favor of German thinkers (7), and M.H. Abrams largely ignores him in his two great books on the movement, which concern the English Romanticism, *The Mirror and the Lamp: Romantic Theory and the Critical Tradition* (Oxford: Oxford UP, 1971) and *Natural Supernaturalism: Tradition and Revolution in Romantic Literature* (New York: Norton, 1973).

CHAPTER 5

Courses of Empire: Ecological Apocalypse in Early American Literature

American Indians

This chapter traces the development of views about the natural world and the human position in it among American writers from the early colonial era to the middle of the nineteenth century, when the still-new nation had established a great literature of its own. These writers Americanized European ideas about ruins, fallen civilizations, and ecological apocalypse, and some of them question anthropocentrism and set the stage for a full-blown ecocentrism. Before proceeding into the works by selected European Americans, I devote a few pages to ideas on the place of humans in the cosmos held by American Indians. Chronologically, this chapter is out of order, since there is not a lot of what is traditionally considered literature by American Indians before the twentieth century. But later American Indian writers inherited worldviews from their ancestors that are rich in myth and respect for the natural world and which are in stark contrast to those of the strongly anthropocentric European colonists, and much the same contrast persists in the twenty-first century.

After Columbus, Indian storytelling and, later, literature, would reflect the overwhelming forces of colonization and struggles for self-identity. It is at once a cliché and not wholly accurate to state that American Indians in the colonial era lived in complete harmony with their environment. They put strains on the natural world, and many of them held at least a soft anthropocentrism, but the Indians' animism, the idea that spirit is found not only in humans but in the nonhuman (plants, animals, bodies of water,

B.L. Moore, *Ecological Literature and the Critique of Anthropocentrism*, DOI 10.1007/978-3-319-60738-2_5

and so on), instilled them with a sense of restraint lacking in most Europeans (A. Taylor 18–19).[1] Later Indian statements about the "ownership" of land, an alien concept in the European sense to pre-Columbian Indians, were delivered within the context of white aggression and colonization. For example, in 1805 the Seneca orator and chief Red Jacket said, "There was a time when our forefathers owned this great island. Their seats extended from the rising to the setting sun. The Great Spirit had made it for the use of Indians. He had created the buffalo, the deer, and other animals for food. He had made the bear and the beaver. Their skins served us for clothing" (230). Yet American Indians generally lacked the European sense of ownership and Baconian will to exploit nature for works in the European sense. The Europeans segregated the natural and the spiritual and tended to see the Indians' animism as devilish. Many Indian people practice an ecological self-sufficiency that has worked for over one thousand years. Modern Hopis and Zunis observe the extravagance of American society all around them and conclude, "We were here long before you came, and we expect still to be here long after you too are gone" (Diamond, *Collapse* 143). The Anasazi civilization survived for about 600 years, which is longer than Europeans have been in North America, while the Pueblo Indians have sustained themselves for 1000 years. These examples should give us pause regarding the sustainability of First World economies (Diamond, *Collapse* 155).

The context for discussing American Indian literature and the anthropocentric treatment of the earth differs markedly from that of Euro-American literature because it is flushed with Eurocentric-induced genocide, a hegemonic element inseparable from anthropocentrism. This is, understandably, an overwhelming context and subject for many Indian writers. Sidner Larson (Gros Ventre) acknowledges the genocide enacted by whites, the near extinction of a people that have survived and carried on, but he believes that it should not be the focus for contemporary American Indian writers: "The discussion related to the genocide perpetrated upon the original inhabitants of the Americas is both absolutely necessary, and crucial to move beyond; it is a primary moving force and the undertow that drags Indian people into a sea of self-hatred time and again" (144).

Bridging the gap between the European interest in the fall of empires with the ravaged context of post-Columbus American Indians, Peter Blue Cloud's (Mohawk) poem "Crazy Horse Monument" asks what Crazy Horse would

. . . think of the cold steel chisel,
and of dynamite blasting a mountain's face,
what value the crumbled glories of Greece and Rome,
to a people made cold and hungry? (Naitum p. 80)

In a *New York Times* editorial, Louise Erdrich (Ojibwe) writes that everyone has tried to envision what the world would look like after a devastating event; for American Indians, it has already happened: "Many Native American cultures were annihilated more thoroughly than even a nuclear disaster might destroy ours, and others live on with the fallout of that destruction, effects as persistent as radiation—poverty, fetal alcohol syndrome, chronic despair" ("Where" 23).

John Elder writes that, unlike Euro-American poetry, Native American poetry is not predicated upon a conflict between humanity and nature but is centered more in a mythic view of nature (712). Yet for American Indians, the concept of wilderness is a European idea that, at least up to the near present, has been used to erase their memory from the land. Says one Indian rights activist, "There was no wilderness until the Whites arrived" (A. Miller 109). Commenting on having witnessed the cavalier and brutal treatment of rainbow trout by two men from the Department of Fish and Wildlife, Linda Hogan (Chickasaw) calls such actions as "far-hearted," a term employed by Bushman people. Such an attitude "is one that sees life, other lives, as containers for our own uses and not as containers in a greater, holier sense." Hogan believes that even wilderness areas are considered valuable only insofar as they serve people:

> While most of us agree that wilderness is necessary to our spiritual and psychological well-being, it is a container of far more, of mystery, of a life apart from ours. It is not only where we go to escape who we have become and what we have done, but it is also part of the natural laws, the workings of a world of beauty and depth we do not yet understand. It is something beyond us, something that does not need our hand in it. As one of our Indian elders has said, there are laws beyond our human laws, and ways above ours. (45)

In some cases, conservationist attempts to restore wilderness areas to their conditions before white men arrived have had negative impacts, as on Havasupai and other American Indians (Johnson 124).

An underlying sense of ecology is inherent in Indian lifestyles. Joseph Bruchac (of Abenaki but also English and Slovak ethnicity) writes in "The Circle Is the Way to See" about American Indian myths and how human overuse of the land leads to destruction and dystopia. In his short poem "Prayer," he writes that if humans pretend we are the center and that mammals, fish, and birds "are at the edge of grace, /then we circle, dead moons/about a cold sun." Regarding the earth's sickness, he writes that humans, who have the unique ability to upset the natural balance, are a big part of the problem: "We are not the strongest of all the beings in Creation. In many ways we are the weakest." We have, he continues, forgotten the instructions from God to "be kind to each other and respect the Earth." Indian stories and myths "help us recognize our place as part of the circle of Creation, not above it." Humans are meant to live in harmony and not "threaten the survival of the insect people or the whale people" ("The Circle" 815).

Bruchac concludes his study of Indian storytelling *Our Stories Remember* with a discussion of the human relations to our fellow animals and plants, and the traditional Indian attitude is clearly ecocentric, even though Indians, especially those living where growing seasons are abbreviated, have a long history of hunting for food. He writes, "Seeing animals as of lesser value than humans has always been called foolishness in American Indian cultures. Not only traditional stories, but personal experience, taught the elders of all our nations that the Animal People care for their families, feel love and sympathy, anger and despair just as human beings do. By observing animals, humans can learn many things" (158). The notion that "Animal People" would have "sympathy" for humans is linked to the unclear distinction between human and animal: "the line between human and animals is so lightly drawn in American Indian cultures that it ceases to exist at certain points" (160). In fact, according to a story by the Ho-Chunk people of the upper Midwest, the Great Spirit decided to give tobacco to humans, not because they are strong but because they are "the weakest and most confused things in Creation" (169). Indians, according to Bruchac, show respect not only to their fellow animals but also to the Earth's plants, which are "sacred and sentient" (164).

Such notions of sentience, interconnectedness, and sympathy infuse the poetry of American Indian poetry from the first half of the twentieth century to the present. In Gail Tremblay's (Mi'kmaq and Onondaga) "Night Gives Old Woman the Word," the sense of the interdependence of things is not contained on the earth, as the moon "makes corn leaves uncurl/and probe nocturnal air" while "Clan mother, watching, /hears the planets move," and in her poem "Medicine Bearer," a man "dances on this spinning planet." A strong sense of the connectedness of things fills Anita Endrezze's (Yaqui and European) poetry as well as the notion that the human view of nature is only one of many: "What the owl flies into/we call night. /The moon is a windfall, a pear/weathering to the core" (Naitum p. 319). In her poem "Don't Forget," A. Sadongei (Kiowa/Tohono O'odham) asks, "When coyote talks to you/do you listen?" Counter to Cartesian dualism, she writes that "spirits long dead/live on in trees and rocks" (Naitum 353). Peter Blue Cloud's "Wolf" is a strong, compassionate poem about the sufferings of wolves, which are hunted by men, though "curious men" can learn from the wolves' "wisdom" (Naitum 92–93). And Louise Erdrich, in her often-anthologized poem "Dear John Wayne," fuses Hollywood images of white colonialism with the Indian concept that the earth cannot be owned. At a drive-in movie, Wayne's patriotic character announces, "*Everything we see belongs to us*" (l. 23), but the reality is, "*Death makes us owners of nothing*" (27).

Much of the poetry of Joy Harjo, discussed in Chap. 6, is filled with the idea that humans are akin to the earth and the stars—the stuff of the universe: "the Earth . . . is one of us . . . she loves the dance for what it is. So does the Sun who calls the Earth beloved" (176). The earth is a holy place, "the feet of god/Disguised as trees" (183). The title of Harjo's collection of poetry *How We Became Human* suggests, following both Indian myth *and* modern science, that humans are derived from the earth, are a part of it. Observing rain on leaves near her residence in Honolulu, she states that humans may quietly "listen to the breathing beneath our breathing." This, she, states, is "how we became human" (194).

In an interview with Bruchac, N. Scott Momaday (Kiowa) said, "The whole world view of the Indian is predicated upon the principle of harmony in the universe. You can't tinker with that; it has the look of an absolute," and he rejects what he sees as a modern tendency toward alienation: "On the basis of my experience, trusting my own perceptions, I don't see any validity in the separation of man and the landscape . . . I think it's an unfortunate point of view and a false one, where the relationship between man and the earth is concerned" (14-15). Elsewhere, Momaday states that the native American view of the natural environment is not scientific and rational but a "more imaginative kind. It is a more comprehensive view. When the native American looks at nature, it isn't with the idea of training a glass upon it, or pushing it away so that he can focus upon it from a distance. In his mind, nature is not something apart from him. He conceives of it, rather, as an element in which he exists" (Capps 84).

A similar idea concerning the traditional view of Indian interrelatedness in conflict with white ownership emerges in Leslie Marmon Silko's (Laguna/Pueblo) well-known novel *Ceremony* (1977), set in the land of Navajo and Pueblo Indians. The novel, though nonlinear, is built largely around the experiences of post-World War II, PTSD-suffering veteran Tayo, who embodies the double-consciousness of being half-white and half Indian. Following the tradition of his ancestors, Tayo covers a deer's head "out of respect" before his cousin Rocky slits its throat, while Tayo's uncles give the deer pinches of cornmeal to feed the deer's spirit. These actions embarrass Rocky, who has in respects integrated himself into white culture (46–47), though he will later die in the Bataan death march, which Tayo will also endure but survive. Though tortured mentally by his war experience, he is intent on ceremonial healing by finding links between himself, his home land, and people.

A part of Tayo's ceremonial healing is the recovery of truth amidst falsehoods of a thoroughly anthropocentric white culture. Native elders, by contrast, Lawrence Buell writes, "traditionally think of the world as a place-centered continuum of human and nonhuman beings" (286). Tayo's "old Grandma" tells him that once "animals could talk to human beings and many magical things still happened," an idea Tayo carries with him despite what he has learned in school (87). Betonie, a medicine man Tayo visits in Gallup, tells him that the white men flaunt the idea that they have stolen the land from them: "They only fool themselves when they think it is theirs. The deeds and papers don't mean a thing. It is the people who belong to the mountain [Mount Taylor]" (118). An Indian witch, says Betonie, created white people, who "grow away from the earth" and the sun, the plants, and animals. As for the white people,

> They see no life
> When they look
> they see only objects.

The whole world, for the whites, is dead, so "[t]hey destroy what they fear" (125). Later in the novel, confirming Betonie's words, a white land owner (Floyd Lee) all but destroys the land in order "to make the land his" (174). Tayo is pursued by the police, who believe he is a lunatic, and he hides in an abandoned uranium mine, the contents of which went into making the atomic bombs dropped on Japan. Near the end of the novel, Tayo beholds a beautiful vista near Enchanted Mesa at dawn and feels, for a moment, "balance" in his mind, but that is enough: "The strength came from here, from this feeling. It had always been there. He stood there with the sun on his face, and he thought maybe he might make it after all" (220–21).

Colonial and Early National Era

Set against such high expectations, the New England Puritans' vision of a "Bible Commonwealth" was doomed to failure. The numerous dystopian jeremiads appearing in New England toward the end of the seventeenth century bespeak a desire to "regain" the utopian goals established by John Winthrop and other New English leaders in the early years of the Plymouth colony. Not symptomatic of a Puritan "declension," the dystopian visions dramatized in the jeremiads were meant to convince Puritan audiences that they needed to work harder to bring about Winthrop's utopian "City upon a Hill" (see A. Taylor 185–86). For Winthrop, the undeveloped land of the Indians was in itself a divine argument for white ownership. He wrote in a 1639 letter that the white settlers took possession of Indians' lands "peaceably . . . being thus taken and possessed as *vacuum domiclium* gives us a sufficient title against all men" (*Winthrop* 101). Winthrop writes in *A Model of Christian Charity* that in order to avoid the "shipwreck" of God's judgement, "we must be willing to abridge ourselves of our superfluities, for the supply of others' necessities" (176), but the sentiment did not apply so much to Indians, whose long-established habitats would over coming decades become raw material for Euro-American enterprise.

William Bradford infamously refers to the wilds of New England as "a hideous and desolate wilderness, full of wild beasts and wild men" (70), an idea echoed by other seventeenth-century colonists, including Mary Rowlandson in her popular captivity narrative of being taken by Indians into a "vast and desolate wilderness" (260). Yet very early in the seventeenth century a number of writers challenged the colonizing, anti-naturist tendencies of the leadership. William Wood published his book of New England natural philosophy, *New England's Prospect* (1634), in his native England, and though not a Puritan, he arrived in New England before Winthrop. As Michael P. Branch writes, Wood states that the flying squirrel is of no human use, "a creature more for sight and wonderment than either pleasure and profit" (45), hence suggesting a nonanthropocentric viewpoint (99). In her poem "Contemplations," Anne Bradstreet marks the first American poetic attempt to examine the cleavage between fallen man and God's nature and showing the impermanence of earthly life with eternal heaven. Unlike most of her contemporaries, her sheer appreciation for God's creation is direct. Her final stanza evokes themes in much of the British poetry discussed in the previous chapter:

> O Time the fatal wrack of mortal things,
> That draws oblivion's curtain over kings;
> Their sumptuous monuments, men know them not,
> Their names without a record are forgot,
> Their parts, their ports, their pomp's all laid in th' dust
> Nor wit nor gold, nor buildings scape times rust;
> But he whose name is graved in the white stone
> Shall last and shine when all of these are gone. (225–32)

Bradstreet's poem is wholly Puritan, but it is also a nature poem and, in certain respects, preromantic in its clear love for nature and in its elegiac tone regarding the brevity of human life. Many of Edward Taylor's poems, including the "Prologue" to his *Preparatory Meditations*, evoke a similar humility, the nothingness of humans. Yet, praising eternal heaven, the poem shows that a crumb of dust (the soul of a single member of the elect) outweighs the Earth because the elect exist for eternity, while the earth will pass away, a reversal similar in Stanza 20 of Bradstreet's "Contemplations." Theocentrism is not all of one sort; some versions of it lean heavily toward a human-centered view of the world, and this view dominates New England Puritanism. Another type of theocentrism, evident in the writings of Bradstreet (and later in those of John Muir and many others), lean more heavily toward humans as a part of God's creation.

A thorough sense of apocalypse as foretold in the Bible is a central in Michael Wigglesworth's long didactic poem *The Day of Doom* (1662), which was the bestselling book, next to the Bible, in seventeenth-century America. The natural environment of the poem is wholly a material backdrop for God's judgement, but in his poem *God's Controversy with New-England*, written, states the subtitle, "in the Time of the Great Drought, Anno 1662," he echoes Bradford in assessing the land as

> A waste and howling wilderness,
> Where none inhabited
> But hellish fiends, and brutish men

That Devils worshiped. (25–28)

Cotton Mather's paper on The Great Comet of 1680 and Halley's Comet of 1682 suggests that these heavenly phenomena are signs of God's judgment of sinners. He quotes Newton and Edmond Halley and then the Scottish physician George Cheyne: "these frightful Bodies are the Ministers of *Divine Justice*" (53). (Cf. Fontenelle's *The Comet*, published almost contemporaneously; see above.) In Jonathan Edwards's conception of order as depicted in "Sinners in the Hands of an Angry God" (1741), human existence is fraught with the danger of displeasing God, who holds humans by a spider's thread over a flame. Where Bradstreet's theocentrism is based on an earth that extols the wonders of God's creation, a beautiful but poor image of the saints' hereafter, in Edwards's famous sermon, "the earth does not willingly yield" to humans/sinners, nor does the sun. Likewise, "God's creatures are good, and were made for men to serve God." Yet, apparently rejecting the Cartesian idea that animals are automatons, animals "groan when they are abused to purposes so directly contrary to their nature and end" (96).

Newton and the new sciences shook up the relatively simple understanding of order in the cosmos, and deism would affect even those who held traditional Christian views. In Phillis Wheatley's poem "Thoughts on the Works of Providence" (1773), written while she was a slave at around the age of twenty, not only creation but science underscore God's glory. For Wheatley, who had some understanding of a heliocentric universe and Newtonian science, God's creation, including a sun that provides life on earth, is full of wonders "by men explored, /And yet creating glory unadorned" (27–28). Even the secret restorative powers of sleep—reason "suspended in nocturnal hours"—is a gift from God that bespeaks his love for us (93–96). A remarkable poem by an imaginative young woman (another of her poems is an ode to imagination itself), the poem nonetheless concludes with a moral—and anthropocentric—tag: all of Nature serves "one gen'ral end, /The good of man" (127–28), though he is little aware of it.

Perhaps the most direct attack on anthropocentrism—and theocentrism—in American letters in the eighteenth century is Thomas Paine's book (delivered originally in a series of pamphlets) *The Age of Reason*. Deeply deistic in arguing that humans can learn about creation through reason, Paine shows skepticism toward institutional Christianity and, along with it, anthropocentrism, sometimes within the context of larger ideas. For example, he writes near the end of Chap. XI:

> The Almighty lecturer, by displaying the principles of science in the structure of the universe, has invited man to study and to imitation. It is as if he had said to the inhabitants of this globe that we call ours, "I have made an earth for man to dwell upon, and I have rendered the starry heavens visible, to teach him science and the arts. He can now provide for his own comfort, AND LEARN FROM MY MUNIFICENCE TO ALL, TO BE KIND TO EACH OTHER."

In the following paragraph, he argues, echoing the French philosophes, that the distant planets and stars—an "immense desert of space glittering with shows"—have little to do with man, though they are useful for contemplation and for showing him about himself (55). Many of the nation's founders were also deists (or influenced heavily by deism) and held similar views, including Jefferson, Madison, and Franklin, whose autobiography, incomplete and unpublished during his lifetime, rejects the cosmic order of natural philosophy for a rational, deistic one.

Two American writers of French heritage who published in the decades before and after the turn of the nineteenth century mark the changing perceptions of the place of humans on earth, mixed here and there with ideas about ruins and decay, and both writers help account for the bridge from the Enlightenment to the romantic era in their appreciation of wild nature, yet they differ in many ways. J. Hector St. John de Crèvecoeur was born in France and immigrated to New France (Quebec) when he was twenty, while Philip Freneau, of Huguenot/Scottish descent, was born in New York and grew up in New Jersey. Crèvecoeur became a British subject, and, a loyalist, would under the yoke of suspicion of spying for the British, leave North America and settle eventually in France. Freneau was a revolutionary—the "Poet of the American Revolution"—and the most committed of Jeffersonian anti-Federalists.

Written while farming in Orange County, New York, Crèvecoeur's *Letters from an American Farmer* (1782) assumes the intelligent, often satirical voice of a farmer-narrator named James in twelve letters addressing a British gentleman about life in the colonies, often contrasting American views with European ones. In Letter I, a minister friend of James compares an English audience, given to admiring "ruins of temples," which imparts "a knowledge which appears useless and trifling," with America, which is filled with the "rudiments and embryos" of societies, not Old World "ruins of old towers, useless aqueducts, or impending battlements" (42). Crèvecoeur returns to the idea at the end of Letter VIII, regarding the antiquities of Italy versus the woods of America, which are more conducive to freedom: "we enjoy in our woods a substantial happiness which the wonders of art cannot communicate" (165). Lest we conclude that Crèvecoeur is an early Thoreau, he much more regularly celebrates the relative wilds of America not in itself but for its newness, its potential for farming in particular. His female copulation metaphor illustrates this —"Nature opens her broad lap to receive the perpetual accession of newcomers and to supply them with food" (42–43)—and on the very next page (as if to make the point clear), he employs, not for the last time, a metaphor of the land as a woman, as American farmers "penetrate deep" into her (see Kolodny 52–63). America has the capacity for individuality, freedom and patriarchal, Baconian wealth-making. One of his feet (or at least a few toes) remain in the Puritan view of the wilds as wicked; he refers to meadows that were at one time "a hideous wilderness, how converted by my industry into rich pastures and pleasant lawns" (216).

Yet Crèvecoeur also bemoans the vanishing Indians of Nantucket, and he is, in Letter IX, out front in railing against slavery. He returns, also near the end, to a discussion of ruins. Nations have disappeared from Nantucket; the Indian population is dwindling, while Nantucket homes are built on the ruins of "ancient huts" (162). Anticipating William Cullen Bryant of "The Prairies" but also invoking the late eighteenth-century fascination with fallen civilizations (and perhaps speaking as a loyalist against the colonial revolutionaries), he writes that the "history of the earth" is composed of "whole nations devoted to the blind fury of tyrants. Countries destroyed, nations alternately buried in the ruins by other nations" (173). Thinking about the wind's prodigious power, so demonstrable in this part of the nation, man appears "diminutive" (164). The idea that the people of the frozen north are superior to the bears they hunt only in their ability to speak is probably less an identification with wildness or statement of the kindredness of living things than a bitter view of humanity, post-Revolution (175). Near the end of Letter XII, he vows to seek solace in the wilds. He writes that the "immense variety of planets" created by God should place us "wretched mortals" in a position of humility (226).

Like Crèvecoeur, Freneau was a deist preoccupied with wild nature, and they both comprise a part of the bridge from American Enlightenment concerns toward a more full-blown American romanticism. Unlike

Crèvecoeur, Freneau was a Jeffersonian, a revolutionary, and then an anti-Federalist voice in the new republic. Early in his career as a poet, he followed the European interest in fallen civilizations. His poem "The Pyramids of Egypt," written in 1769 while still a teenager, is a dialogue between an inquiring "Traveller" to the nation's ruins and "Genius," the attendant spirit of Egypt. The former remarks on the wonders of Egypt:

> Since these proud fabricks to the heavens were rais'd
> How many generations have decay'd,
> How many monarchies to ruin pass'd!
> How many empires had their rise and fall!
> While these remain—and promise to remain . . . (p. 204)

Genius replies that someday, these wonders "shall moulder on their bases, /And down, down, low to endless ruin verging, /O'erwhelm'd by dust, be seen and known no more!" Genius' rhetorical question—"where is Pharaoh's palace, where the domes/Of Egypt's haughty lords?" (205)—anticipates Shelley's "Ozymandias" by almost fifty years. The patiently apocalyptic force of Time personified closes the poem with a rundown of other ancient monuments it has razed. The Earth, too, will follow suit.

Freneau would turn his apocalyptic vision to America in such poems as "The American Village," which is an American reply to Oliver Goldsmith's "The Deserted Village." Like Crèvecoeur, Freneau's American pastoral was a safe harbor from the tyranny and violence of the Old World, but should America someday follow Europe into overdevelopment, the result will be its own destruction:

> Like ROME she conquer'd, but by ROME she dy'd:
> But if AMERICA, by this decay,
> The world itself must fall as well as she.
> No other regions latent yet remain,
> This spacious globe has been research'd in vain.
> Round it's whole circle oft' have navies gone,
> And found but sea or lands already known.
> When she has seen her empires, cities, kings,
> Time must begin to flap his weary wings;
> The earth itself to brighter days aspire,
> And wish to feel the purifying fire. (218–19)

His most famous poem, "The Wild Honey Suckle," published in 1786, romanticizes decay, well before Bryant, Byron, and Wordsworth. Like many of his poems after the Revolution, it is about the transiency of life, the inevitability of decay. Freneau's deism led him to poeticize the natural laws of life and death, and objects in nature show us how to accept life's brevity. Nature reflects God's design and is thus worthy of celebration. Against the transiency of life, civilization is a folly that smacks of tyranny, while nature is simple and the true center of existence.

Now more forgotten than Crèvecoeur and Freneau, Timothy Dwight wrote a long poem entitled *Greenfield Hill* as a tribute to Connecticut. In florid language, the poem celebrates the natural beauty of the state, but it is also heavily didactic and sentimental. Part IV, "The Destruction of the Pequods" (1794), reflects on the near extermination of the Pequot people by English colonists in the seventeenth century. The opening stanza ponders "the long vale of time" on earth:

> AH me! while up the long, long vale of time,
> Reflection wanders towards the eternal vast,
> How starts the eye, at many a change sublime,
> Unbosom'd dimly by the ages pass'd!
> What Mausoleums crowd the mournful waste!
> The tombs of empires fallen! and nations gone!
> Each, once inscrib'd, in gold, with "AYE TO LAST,"
> Sate as a queen; proclaim'd the world her own,
> And proudly cried, "By me no sorrows shall be known." (5–13)

The remainder of the poem is a roundabout account of how the English colonies, headed by Captain John Mason and powered by the colonial army's superior weaponry, massacred the Pequot and drove survivors from their ancestral home. Yet Dwight is optimistic about the new republic, which is, for him, a sort of utopia. Superior to Europe in its agrarian economy, he little considers that white society could ever go the way of the American Indian.

Early American Romanticism

The temporality of European culture in America is a central idea in much of the work of William Cullen Bryant, for a time the new nation's greatest poet. Though known essentially as a nature poet, Bryant would become a city dweller and a man of wealth through editing the *New York Evening Post* for almost half a century. He would speak out later in life for land preservation in his campaign for Central Park in the 1840s, but by this time he had long been taken for granted as a poet. Although his poetry is now largely overlooked, many of William Cullen Bryant's poems treat the old idea of fallen empires and races with more deeply romantic nature imagery, inspired by the English romantics and eighteenth-century Graveyard poets, to develop a unique if inconsistent body of work. In his 1821 poem "A Walk at Sunset," the speaker weighs the immutability of nature against changing humanity—a common Bryant theme. He evokes various images of the landscape, "the glories of the dying day/Its thousand trembling lights and changing hues" (23–24), then considers the fact that the same sun shone "before the red man came" and (believing a race populated the region before the American Indians) looked down on "the hunter tribes" (32, 37). But "Now they are gone" (43), and as the speaker beholds the landscape, he acknowledges that the sun "Must shine on other changes" (56). Although the poem is not a direct comment on the over-industriousness of the white man, it does envision a day when nature will continue to exist long after white culture is gone.

The idea of the earth as a giant grave is a common theme in Bryant's poetry. The title character of his "An Indian at the Burial Place of His Fathers" looks on the past and mourns his vanishing race, but he begins to think of the "pale race" of the future: "Their race may vanish hence, like mine/And leave no trace behind/Save ruins" and "white stones above the dead" (63–66). In his most famous poem, "Thanatopsis," the first version of which Bryant wrote as a very young man, when we die, we are buried in a beautiful earth; its hills, vales, woods, rivers, brooks, meadows, and oceans "[a]re but the solemn decorations all/Of the great tomb of man" (37–45). We should approach death as "one who wraps the drapery of his couch/About him, and lies down to pleasant dreams" (91–92). And in his poem "Earth," he apostrophizes a female earth: while reclining alone on her soil, "[t]he mighty nourisher and burial-place/Of man, I feel that I embrace their dust" (39–40).

When Bryant visited his brothers near Jacksonville, Illinois in 1832, he did not think the prairies particularly beautiful, but the visit did inspire one of his best and best-known poems, "The Prairies," which is one of the first (non-American Indian) literary works celebrating that region of the country as a unique American treasure (see C. Brown 205–07). Written in blank verse, the poem draws on many of the themes that Bryant had explored over the years in other poems. One theme is the superiority of the undeveloped American landscape over the long-spoiled, overdeveloped European world. Bryant establishes this idea immediately as the speaker notes that England has no name for prairies because it has no such lands (3–4). Echoing his poem for the Europe-bound Thomas Cole, in which he states that everywhere in Europe is "the trace of men" ("To Cole" 10), here, the speaker states, regarding the prairie, that "Man hath no part in all this glorious work" (24). Thus, for Bryant as for Cole, undeveloped land is superior to that which is developed. The prairie, Bryant suggests, has seen cultures as old and perhaps as advanced as Europe's. The speaker is not encouraged by his imagined future development of the prairie, where a busy beaver "rears his little Venice" (97)—Venice functioning as a synecdoche for the corrupt culture of Europe. As the settlers make their way west to the prairies, he compares them to a bee, another European import:

> . . . I listen long
> To his domestic hum, and think I hear
> The sound of that advancing multitude
> Which soon shall fill these deserts . . . (104–07)

In effect, Bryant fears that the prairies, which have for the moment been reclaimed by nature from its former inhabitants, will, thanks to what would soon be called Manifest Destiny, before long look a lot like Europe, and doubtless, the marks left by the European inhabitants will be more permanent than those of the mound-builders or the Indians.

The speaker of the poem, as in others, regards the earth as a giant grave. As his horse walks through the prairie grass, the poet is mindful "of those/Upon whose rest he tramples" (38–39), and he sees the mounds —"the piles of earth that hold their bones" (66). Bryant is disturbed that culture, represented by himself and his horse, tramples "sacrilegiously" on history, that progress has so short of a memory. Attached to the idea of the earth as a grave is Bryant's poetic impulse to contrast the unchanging earth with ever-changing human cultures. In his cyclical view of history, one civilization thrives and then decays and makes way for another civilization that undergoes the same process. In his cataloguing of the civilizations that lived on the prairie, the question arises (one that Bryant doesn't try to answer): is European culture here to stay in America, or will it, too, leave only remnants of its existence? The poem also examines the key Bryant idea of nature as a teacher of spiritual lessons, especially toward the end of the first of the three long stanzas. As he would write in another poem, "The groves were God's first temples" ("A Forest Hymn" 1), so is the prairie a "magnificent temple of the sky" (29), and the heavens "Seem to stoop down upon the scene in love" (32). Though the lands had been used by humans and are in the process of being used again, the hordes being on their way to bring New England to the Illinois, nature is still innocent and sacred, capable of showing eternal truths to those willing to look and listen.[2]

Where Bryant demonstrated that a poet born in America could write poetry of interest to Europeans, James Fenimore Cooper demonstrated much the same in novel-writing. Like many of Bryant's poems, some of his *Leatherstocking Tales* show interest in the late Enlightenment idea of the historical cycles societies and nations. His *The Last of the Mohicans* (1826) suggests that the new nation, only half a century old, is also subject to the rise and decline, though the plot is based in the year 1757, during the French and Indian War. Near the end of the novel, Chingachgook, Mohican chief and Cooper hero Natty Bumppo's companion in most novels in the series, says, "I am a blazed pine, in a clearing of the palefaces . . . I am alone" (877). Chingachgook is literally the title character and the last of his people: he will outlive his sons and die as an old man in *The Pioneers*. White people chop down trees (a recurring cue of white destruction in the novels) and exterminate races, but they will one day face a similar fate due to their wantonness. Bumppo is not only sympathetic to wild nature but is engaged deeply with it, and he is more than a little wild himself. In *The Pioneers*, Bumppo impresses his fellow hunters with his superior shooting skills, but he uses the display to warn the development-minded Judge Temple against wasting nature with heartfelt belief in the inherent worth of nonhuman life:

> "Put an ind, Judge, to your clearings. An't the woods his work as well as the pigeons? Use but don't waste. Wasn't the woods made for the beasts and birds to harbour in? and when man wanted their flesh, their skins, or their feathers, there's the place to seek them. But I'll go to the hut with my own game, for I wouldn't touch one of the harmless things that kiver the ground here, looking up with their eyes at me, as if they only wanted tongues to say their thoughts." (250)

Bumppo's preservationist rhetoric is strong here, as it is in the other Leatherstocking novels, but Cooper's own views were more complex. As

Roderick Nash writes, "Cooper knew that civilization also had its claims and that ultimately they must prevail. The elimination of wilderness was tragic, but it was a necessary tragedy; civilization was the greater good" (77).

More explicitly preservationist than her famous father, Susan Fenimore Cooper's *Rural Hours* (1850/1887) also bemoans the wanton chopping down of trees, but her preservationist ideas are typically implicit and are not bitter. It takes only a few short minutes to chop down a tree and can be accomplished by almost anyone, she notes, whereas a tree takes many years to grow (134). Trees that are not chopped bear witness to many changes in the land, including human works that over time are run over by wild flora and fauna:

> This little town itself [in rural New York] must fall to decay and ruin; its streets must become choked with bushes and brambles; the farms of the valley must be anew buried within the shades of a wilderness; the wild deer and the wolf and the bear must return from beyond the great lakes; the bones of the savage men buried under our feet must arise and move again in the chase, ere trees like those, with the spirit of the forest in every line, can stand on the same ground in wild dignity of form like those old pines now looking down upon our homes. (135)

A major theme in the book is the idea that the human time on the land is very brief, while the forests and trees are long-lasting and have at least an equal value to civilization: "the works of man are ever varying their aspect; his towns and his fields alike reflect unstable opinions, the fickle wills and fancies of each passing generation; but the forests on his borders remain to-day the same they were ages of years since" (142). Cooper is aware of the potential of greed to destroy wild nature, which she notes is a great part of the wealth any civilization possesses (153). Throughout the book, she argues that God's hand is apparent in his creation, but, echoing Cole (and anticipating *Walden* by a few years), the landscape will lose this aspect "if ever cupidity and the haste to grow rich shall destroy the forest entirely, and leave these hills to posterity, bald and bare . . ." The land without its trees are like Samson shorn of his locks (160). A few pages later she bemoans a grove that has been "recklessly abused by kindling fires . . . oaks that might have stood yet for centuries, with increasing beauty, have been wantonly destroyed," and she notes that these trees have disappeared over the years (167). In autumn, she writes that misguided seekers of treasure killed a “singular tree”; “they threw out so much earth, that the next winter the tree died” (248).

Cooper early on establishes the personhood of the nature around her neck of the woods in New York, sometimes mixing an implicit critique of the wanton cutting down of beloved trees: “several noble pines, old friends and favorites, had been felled” (4). Similarly, she gives several birds semi-human qualities; orioles are hard-working, harmless, innocent, possessed of excellent character, affectionate, and grieving (18). Anticipating the first sentence of Aldo Leopold’s *A Sand County Almanac* (“There are some who can live without wild things, and some who cannot” (vii)), she says that nature “gives pleasure to those who are content to await the natural order of things” (10). And like Goethe and Emerson, Cooper strongly believes in the unity of nature, often noting in different contexts its interrelationships, although, she writes, “nature is the common name of all” (61). Yet she also occasionally betrays blind spots common in the mid-nineteenth century. With no explanation, she refers to an Indian as a “savage” (63); in passing, she offers the idea of an “ignorant negro” (248); and, seemingly out of step with her stated love for the forests, she is oddly concerned with improving the market value of timber (248).

Cooper’s narrative is infused with an awareness of the passage of time and the reality of decay. The American Indians were, she notes, as possessed of life as much as white Americans are at present. They drank from the same springs that we today call our own (63). She discusses “the few humble antiquities of our neighborhood”; the ruins of a bridge are not comparable to those of the Old World. American ruins, she writes, “are more rude than theirs” (126). The red man was for a long time the lord of the land but is now forgotten, while “those calm old trees seem to heave the sigh of companionless age” (132).

A few years later, in 1854, Lydia Huntley Sigourney published “Fallen Forests,” which, like Cooper’s *Rural Hours*, castigates man and praises nature in showing how God cares for the simplest things in creation. Shortsighted men chop and burn down forests without a thought for the trees and animals that depend on them or for the possibility that they may provide sustenance for people. The trees in the forest are “nurtured to nobility” (4) and wear their “summer coronets” (5); but unthinking man sets them on fire for “[a]n arch of brilliance for a single night” (10), scaring away the animals who depend on the tree for shelter and food. Lifting “his puny arm” with his axe (14), man chops down the trees, “sacred groves” (17), driving away the birds who live in the trees and conduct “their mute worship”

(22). A rich man builds his huge house, an "ant-heap dwelling . . . but neither he, /Nor yet his children's children, shall behold/What he hath swept away" (42–44). A similar ethos would, a few decades later, characterize many of the chief works of the local colorist Sarah Orne Jewett, including "A White Heron" and *The Country of the Pointed Firs*, both of which suggest that God cares for the small things in nature and that humans are a part of the natural world.

Thomas Cole

Thomas Cole was born in England and emigrated with his family to the United States when he was still a teenager in 1818. He absorbed the insights of Scottish moralists such as Archibald Alison, whose associationist theories on art also influenced Cole's strong sense of nationalism and his religious thinking. He would make art intended to help Americans appreciate the nation's beautiful scenery and converse with God through nature (Baigell 13). Cole's art reflects his belief that all societies pass through material stages. Artistically, his hero was seventeenth-century French landscape painter Claude Lorrain, whose work inspired Cole to import European notions of the Sublime and the Picturesque to his adopted country. Robert Hughes writes, "The idea of landscape, as distinct from mere territory, was imported from England and it appeared quite late in America; Thomas Cole, an English import himself, was its first bearer in painting" (142). Cole had a deeply religious response to his adopted country's embattled wilderness. His family was from Lancashire, which felt the full force of the industrial revolution's blight in England's rural districts (A. Miller 93). Cole believed it a sin to clear forests for farming, an idea observable in his late painting *Home in the Woods* (1847), which portrays a family that sustains itself through fishing and hunting alone (Stoll 67). For Cole, wilderness was an "extended genesis. The wilderness revealed the work and hand of God" (Baigell 10).

After studying at the Pennsylvania Academy of Fine Arts, he moved to New York in 1825, met Bryant and Cooper, and, after journeying up the Hudson River to the Catskills and Highlands, sketched paintings for a suite of paintings that would launch his career and the Hudson River School, members of which would also include Asher Durand, Albert Bierstadt, and Frederick Church, among others (Baigell 10). Like other romantics, Cole's views on American wilderness are complex. His patrons were wealthy Federalists wary of democracy and Andrew Jackson and who held nostalgia for a pristine America. In Cole, the patrons found a kindred spirit; nostalgia was also central to his imagination. Like other American Federalists, Cole identified "with the early Roman Republic whose form of government, architecture, and sculpture" they embraced (L. Ferber 190). Many of Cole's patrons were having their cake and eating it too; even as wealthy patrons such as Robert Gilmor, Jr. preferred paintings with "unmolested nature," business interests were destroying nature (Baigell 22).

For the average farmer, the concept of "landscape" was all but nonexistent; land was not for aesthetic appreciation or spiritual insight but a raw material for exploitation (Hughes 141–42). Cole's view of wilderness has much more in common with James Fenimore Cooper, Washington Irving, and Francis Parkman than with later American ecologically-minded writers such as Thoreau, Muir, or Leopold. Irving's "The Legend of Sleepy Hollow" revolves around the voracious Ichabod Crane, for whom land, like the food prepared for him by the Tarry Town locals, is a thing to be devoured but not loved. Amidst the beautiful natural setting of upstate New York, Crane, reflecting industrial interests, feels the strong need to consume (1067, 1074–75). Two of Cooper's 1820s novels, *The Spy* and *The Pioneers*, understood democracy as dangerous to orderly government and the natural environment. Cole's suspicion about such progress is an underlying theme in much of his work.[3] Roderick Nash notes that Cole "at times . . . dreaded the wilds. The line between the sublime's delightful horror and genuine terror was thin." Yet his "Essay on American Scenery," is the "most distinctive, and perhaps the most impressive, characteristic of American scenery is its wildness" (Nash 79).

In the essay, Cole writes that the liberal arts "soften our manners" and "mend our hearts" (1). He links religion with American scenery in particular, often with special attention to the Hudson River and Catskills (2–4). Though his friends Cooper and Bryant (as well as Irving) made similar statements around the time, Cole is forward looking (anticipating Thoreau by a decade or so) in his disapproval of the diminishment of the American wilds, of a "meager utilitarianism" that is "ready to absorb every feeling and sentiment, and what is sometimes called improvement in its march" and which threaten to crush the imagination "beneath its iron tramp." He urges Americans to "cultivate the oasis that yet remains to us, and thus preserve the germs of a future and purer system." There are things one can learn only from wild nature, goods unattainable in the market (2). Cole observes that "the ravages of the axe are daily increasing—the most noble scenes are made desolate, and oftentimes with a wantonness and barbarism scarcely credible in a civilized nation" (7). We are hardly aware that we Americans "are still in Eden" (8). He knows that the "cultivated state" of Europe is fast approaching his adopted continent,

> but nature is still predominant, and there are those who regret that with the improvements of cultivation the sublimity of the wilderness should pass away: for those scenes of solitude from which the hand of nature has never been lifted, affect the mind with a more deep toned emotion than aught which the hand of man has touched. Amid them the consequent associations are of God the creator—they are his undefiled works, and the mind is cast into the contemplation of eternal things. (3)

Cole writes that it is the sense of solitude, pure works of God untouched by man, that are at the core of America's unique beauties; contrasted with the sublimity of the ruins of Rome, the "American antiquity" where "the sublimity of a shoreless ocean un-islanded by the recorded deeds of man" is more "venerable" (7). In the essay, mountains are the most conspicuous objects in the landscape, symbols of "external majesty, immutability and repose"—characteristics not attributed to the human form or civilization in Cole's paintings.[4]

Cole's grand statement on the temporality of empire mixed with a critique of Jacksonian industrial development is his five-painting series *The Course of Empire*, which, much like a literary work, represents ecological apocalypse, one much more pessimistic than Bryant's, and more didactic. Cole's title and inspiration for the paintings is from George Berkeley's poem "Verses on the Prospect of Planting Arts and Learning in America," the last stanza of which states,

> Westward the course of empire takes its way;
> The first four Acts already past,
> A fifth shall close the Drama with the day;
> Time's noblest offspring is the last. (21–24)

Cole, unlike Bryant a Federalist, delineates a nostalgia for pristine, natural America then being whittled away by Andrew Jackson's expansionist, utilitarian democracy. (In contrast to today and the past century and a half or so, in the early nineteenth century, it was largely the conservative, moneyed, Federalists who were the preservationists.) The paintings are also influenced by the Industrial Revolution, the French Revolution, the Napoleonic wars, and numerous artistic and literary influences, including Lorrain, J.M.W. Turner, Salvator Rosa, Edward

Gibbon, and, not surprisingly, Lord Byron, whose verse, from *Childe Harold's Pilgrimage*, Cole used in his own publicity for the series:

> There is the moral of all human tales;
> 'Tis but the same rehearsal of the past,
> First Freedom, and then Glory—when that fails,
> Wealth, vice, corruption,—barbarism at last.
> And History, with all her volumes vast,
> Hath but one page . . . (4.108.964-69)

The excerpt provides a running narrative commentary for the series. Conceived while visiting the ruins in Rome, the project is rooted deeply in European culture and is a pessimistic view of the rise and fall of civilizations, America in particular (Wilton and Barringer 96).

The first painting in the series, *The Savage State* (1836), introduces viewers to the landscape in which the series narrative unfolds, with the fictional, immutable mountain in background which is positioned at various points in the series. Pink clouds suggest an early hour of day, a beginning, though other elements in the painting (a savage figure on the left and the dark, billowing clouds) conjure "an atmosphere of dangerous exhilaration, with nature red in tooth and claw" (Wilton and Barringer 98). In this, like most of his landscapes, Cole reduces humans to ant-like proportions. *The Pastoral or Arcadian State* (1834) moves Cole's narrative from the Sublime to the "Beautiful." According to Cole, "The simple or Arcadian State, represents the scene after ages have passed. The gradual advancement of society has wrought a change on its aspect. The 'untracked and rude' has been tamed and softened. Shepherds are tending their flocks, the ploughman with his oxen is upturning the soil, and commerce begins to stretch her wings." The rural culture of the painting is reminiscent of Virgil's pastorals, though a Stonehenge structure in the painting may relate the vista to Cole's native Britain. The painting is filled with classical allusions, such as the Clotho (the spinner of fate) figure (Bedell 40). Unlike *The Savage State*, this painting suggests aggression—a soldier, rams clashing, and a tree stump on extreme right indicating an encroachment on the wilderness (Wilton and Barringer 100).

The third painting, *The Consummation of Empire* (1835–36), the centerpiece of the series, "presents a vivid panorama of the vainglorious maturity of a powerful empire" (Wilton and Barringer 102). Amid Romanesque architecture and statues, the eye is drawn to a conqueror in the left foreground robed in red—likely a representation of Jackson—returning amid fanfare toward a triumphal arch. At the point of triumph, seeds of destruction are sewn. The whole scene is characterized by pomp, hubris, and an ostentatious show of wealth. The abundance of human figures suggest overcrowding, and very little wild nature remains. Consequently, *Destruction* (1836) follows. Cole writes, "Luxury has weakened and debased. A savage enemy has entered the city" (qtd. in Wilton and Barringer 105). A storm rages as a city lies in ruins. Finally, the vista is in a state of *Desolation* (1836; see book cover for a detail) resembling the ruins of Rome. "Violence and time have crumbled the works of man, and art is again resolving into elemental nature ... the empire is extinct"; the moon "ascends into the twilight sky near where the sun rose in the first picture" (Cole qtd. in Wilton and Barringer 108). Humans have vanished, and wild nature, represented by the herons nesting on the columns, the abundance of vegetation, and animals grazing without the fear of hunters, has reclaimed the region.

Cole's paintings apply old European ideas about the rise and fall of civilizations and European artistic techniques as a rhetorical response to Jacksonian democracy. As David Schuyler notes, Cole's response to Jackson draws "upon the rhetoric of the eighteenth-century commonwealth ideology in England and America, . . . a throwback to the fear of luxury and centralized power so widely shared by members of the Revolutionary War generation" (40). Observing the rapid development in the northeast, he feared that the business aristocrats were hell-bent on destroying the countryside and that the nation would soon become like Europe (Baigell 19). The nation was indeed changing, and Cole's vision and that of old money New York would give way to an almost unbridled development and Manifest Destiny.

Cole traveled to Europe for two extended periods (1830–32 and 1841–42), especially England and Italy, where he continued to create representations of wilderness in which human dimensions are reduced greatly. His *Italian Scene, Composition* (1833) is, Linda Ferber writes, an "inventory of Grand Tour elements . . . Roman ruins, roadside shrines, dancing peasants, Mediterranean light, and umbrella pines" (168). He was especially moved by the romantic view of Mount Aetna (Etna), which he sketched and painted several times. In the most celebrated of these, *A View from Mount Etna from Taormina, Sicily* (1843), wisps of the volcano's smoke float in background, and a cloaked figure walks through ruins that dominate the foreground; in the middle section lie rolling hills with few signs of human civilization, except for a cityscape partially (and perhaps meaningfully) obscured by the ruins in the foreground. It appears as though the volcano has recently blasted the few human habitations to rubble through which the sole cloaked figure walks.

In his poetry, Cole expresses ideas that complement his art and essays, though, as Marshall B. Tymn (who edited a collection of Cole's verse) writes, his poetry is not distinguished and was probably not intended for close scrutiny (23). Cole developed a love and appreciation for nature by reading Byron and Wordsworth while in England (15). For Cole, the ephemeral beauties of nature are a direct reflection of heavenly eternal truths. Earth, he writes, "is the vestibule of Heaven" (84). Beauties in nature are "glimpses of the world above" by which we "on this darkling sphere" can realize hope (140). Cole states this idea—heaven glimpsed in nature—repeatedly in his poetry, early and late.

His blank verse poem "The Complaint of the Forest" (1838), which was revised by Bryant and published in 1841 as "The Lament of the Forest," is one of the painter's more substantial poems, and it functions as a poetic counterpart to his *Course of Empire* series, *Destruction* in particular. The speaker finds himself at a beautiful lake surrounded by mountains. The sublime ruins of Italy are found wanting compared with this natural vista. The area, a lake "amphitheatre," has experienced its own catastrophe—Cole was an advocate of Cuvier's theories of catastrophes—but it is a "ruin more sublime than if a thousand/Roman colloseums had been pil'd in one" (p. 101). The ancient forest speaks or "complains" to the poet. In olden days, the environment was peaceful. Then man "[a]rose—he who now vaunts antiquity— /He the destroyer." Before men arrived, the trees had echoed only the thunder, winds, water, and birds, but then the "dissonant . . . axe" tore through the forest like a "fierce tornado," leaving ancient mother earth "blasted and bare/Beneath the burning sun" (103). Although, to the benefit of the trees, wars trim the human population on occasion, people invariably return. Yet a "bright virgin continent" (North America) very far away from Europe is populated only by "our native child whose foot/Disdains the sunbeat soil." The white man arrived, few at first, "but soon/The work of desolation was begun" (105), and the "human hurricane—which hath no bounds" appeared. The forest, continuing its complaint, says its "doom is near," as smoke darkens the sky east and west,

> For every valley is an altar made,
> Where unto Mammon and to all the gods
> Of man's idolatry, the victims we are.

In a single day, the forest "growth of centuries is consumed." The land is denuded, and its animal inhabitants lose their homes. The forest's prosopopoeia concluded, the poem's speaker can only cry (106).

The Transcendentalists

Since the New England Transcendentalists believed in the inborn divinity of humans and that God works through humans and through all of nature, it would not be accurate to state that they were explicitly antianthropocentric. Transcendentalism was a syncretic movement, and the divinity of humanity is at once a holdover from mainstream Unitarianism and orthodox Christianity and an appropriation of ancient Stoicism and Eastern philosophies and religions distilled from wide reading. The world, for Emerson and Thoreau, *is* in a sense human centered; as Emerson writes in "The American Scholar," "The world is nothing, the man is all; in yourself is the law of all nature . . . in yourself slumbers the whole of Reason" (51). But this and other seemingly anthropocentric statements, considered in context, which more than anything seek to demonstrate the unity of all things, human and nonhuman, clearly do not function as warrants for a hard anthropocentrism. Emerson and Thoreau, as well as Margaret Fuller (who often focuses on the unity of genders) and Walt Whitman (who almost invariably focuses on the unifying force of democracy), endorse and demonstrate a position of humility with respect to humans in the cosmos. Practically speaking, it may be, ultimately, our humility before the rest of nature that counts the most, beyond purely scientific or doctrinal surety. The Transcendentalists are, at once, anthropocentric in their belief that humans are the godlike end of nature, but that anthropocentrism is soft in the realization that humans do not in any way that matters own the earth, that wild nature is crucial for human progress, and that the Euro-American moment on earth is not permanent and was preceded by other advanced peoples in the past.

As his essays demonstrate, Emerson was well read in the science of his day. Laura Dassow Walls writes that he "took scientific literacy so much for granted that his scientific metaphors sink out of sight" (13). He was aware as any writer so far mentioned of the smallness of humans in the cosmos and the temporality of our species, our institutions and governments. This idea is reflected in major essays and poems, as well as minor ones, such as his poem "The Snow-Storm," which uses a series of images suggesting that nature's artistic powers are superior to human ones. Snow is an isolating element; it blinds the shepherd but also illumines a person who patiently watches the snow as if it were a minister preaching about the underlying unity of all things. His more mature poem "Hamatreya," which Americanizes a passage in the ancient Hindu *Vishnu Purana*, echoes this idea. The poem begins with a listing of men who owned a tract of land and then asks,

> Where are these men? Asleep beneath their grounds:
> And strangers, fond as they, their furrows plough.
> Earth laughs in flowers, to see her boastful boys
> Earth-proud, proud of the earth which is not theirs. (11–14)

Emerson suggests similar ideas about mutability in his essays, including *Nature*. Writing about the relation of nature to language, he evokes Heraclitus: "Who looks upon a river in a meditative hour, and is not reminded of the flux of all things?" (13). And in "Self-Reliance," he dismisses the craze for traveling to exotic locations for self-improvement: "He who travels to be amused, or to get somewhat which he does not carry, travels away from himself, and grown old even in youth among old things. In Thebes, in Palmyra, his will and mind have become old and dilapidated as they. He carries ruins to ruins" (147). Emerson echoes and in a sense universalizes Byron, whose Childe Harold sees himself as a "ruin amidst ruins" (4.25), and he anticipates Thoreau, who states wryly but astutely in *Walden* that he has "travelled a good deal in Concord" (4). Rather than seeking insight at exotic locations, one is better off remaining close to home and cultivating one's own brain, a "few cubic feet of flesh," as Thoreau puts it (5).

Several passages in *Walden* explore the balance between human impermanence, Eurocentric dominance, and the place of humans in the world. The book is certainly one of the greatest affirmations of and guides to self-cultivation, but an understanding of human limitations and the value of other, nonhuman life pervades the book and is essential to that self-cultivation. As in "Hamatreya," observation of nature shows that the human ownership of the land is at best a short-term deal. In "House-Warming" Thoreau imagines a natural world free of Euro-American encroachment: chestnut woods lie in "sleep" under the Concord railroad, ready to reappear when the railroad disappears. The arrogance of Euro-American culture represented by "fatted cattle and waving grainfields" has muted the natural world, but, he writes, "let wild

Nature reign here once more, and the tender and luxurious English grains will probably disappear before a myriad of foes" (238–39).

While hoeing beans (in the chapter "The Bean-Field"), he discovers that "an extinct nation had anciently dwelt here and planted corn and beans ere white men came to clear the land, and so, to some extent, had exhausted the soil for this very crop" (156). But long before the Indians, the woodchuck (a book motif for wildness) had, in the place where Thoreau is hoeing, held "an ancient herb garden" (155). A few pages later, he is aware that his hoeing has "disturbed the ashes of unchronicled nations who in primeval years lived under these heavens" (158). Turning to the trope again, he pauses his work to hear the big guns sounding in Concord in celebration at a time when President Polk is at war with Mexico, and he feels a wave of patriotism and returns to his hoeing "cheerfully with a calm trust in the future" (160). Is Thoreau, here, being sarcastic in his patriotism, or is he thinking about a future in which his militaristic nation no longer controls Massachusetts? He has just quoted Virgil, at once Rome's greatest celebrator of country living as well as its greatest Republican poet. Thoreau concludes that he "will not plant beans and corn with so much industry another summer, but such seeds, if the seed is not lost, as sincerity, truth, simplicity, faith, innocence, and the like, and see if they will not grow in this soil" (164). We are, Thoreau suggests, too anthropocentric and Eurocentric: "the sun looks on our cultivated fields and on the prairies and forests without distinction" (166). He understands that his beans are not just for himself but for the woodchucks, and the squirrels have their own point of view.

In the next chapter, "The Ponds," Thoreau notices the change of time through his observations of a changing landscape, with and without humans, at almost every turn as he explores the ponds scattered through the woods. Such passages look forward the first section of Leopold's *A Sand County Almanac*, which provides a heavily-personified nonhuman history of the land around his Wisconsin farm (see B. Moore 160–69). Thoreau observes that at a section of the pond "[t]here are few traces of man's hand to be seen. The water laves the shore as it did a thousand years ago" (186). He meets an old man who recalls when Walden Pond was "dark with surrounding forests" (190). When he was younger and first floated in a boat on the pond, its shore hills formed "an amphitheatre for some kind of sylvan spectacle" (191), suggesting that the pond was a spectacle outside of human engagement. He notes that woodchoppers are laying waste to trees around the pond, yet "where a forest was cut down last winter another is springing up by its shore as lustily as ever" (192–93). Humans, unlike other animals, require technologies to survive, especially during winter, and through them we are removed from nature, though Thoreau romanticizes his own simple technologies and pared-down living conditions. He turns to thoughts of the extinction of man, in context:

> But the most luxuriously housed has little to boast of in this respect, nor need we trouble ourselves to speculate how the human race may be at last destroyed. It would be easy to cut their threads any time with a little sharper blast from the north. We go on dating from Cold Fridays and Great Snows; but a little colder Friday, or greater snow, would put a period to man's existence on the globe. (254)

Here as elsewhere, Thoreau's language is heavily figurative—humans might as easily expire due to global warming as global cooling—but he is as aware as any writer of the fragility of life, including human.

As did many other writers in the still-new republic, although he is critical of American imperialism and materialism, Thoreau extolls the beauties and superiority of America, which still, in the mid-nineteenth century, held a plenitude of nature. With one eye toward the long-civilized Old World, filled with the ruins of fallen empires, in "Winter Visitors," having few visitors and little to do, he thinks about people, including several African Americans, who once inhabited the area and dealt with its harsh winters, and he wonders why Concord still stands while a village near the pond failed. He writes, "I am not aware that any man has ever built on the spot which I occupy. Deliver me from a city built on the site of a more ancient city, whose materials are ruins, whose gardens cemeteries. The soil is blanched and accursed there, and before that becomes necessary the earth itself will be destroyed. With such reminiscences I repeopled the woods and lulled myself asleep" (264). He prefers a cabin built on wildness rather than civilization. In the following chapter, "Winter Animals," he is aware of foxes running in the snow near his cabin some nights, and he asks, "if we take the ages into our account, may there not be a civilization going on among brutes as well as men? They seemed to me to be rudimental, burrowing men, still standing on their defence, awaiting their transformation" (273). The adjective "burrowing" is, perhaps, not a throwaway word. Suggesting kinship between humans and animals, as well as his own wish for a wilder experience, in the famous, concluding passage of "What I Lived For," he writes, "My instinct tells me that my head is an organ for burrowing, as some creatures use their snout and fore-paws" (98). Concluding "Winter Animals," he asks, "What is a country without rabbits and partridges?" They are "simple and indigenous," "ancient and venerable" and "still sure thrive, like true natives of the soil, whatever revolutions occur" (281). In a day of westward expansion, technological advance (realized most prominently in Concord with the railroad), and faraway wars waged over dubious profit for few and bondage for many, Thoreau is concerned with what is more permanent and of true value.

With the return of spring, Thoreau again turns an eye toward the Old World, dismisses the value of human institutions, and proclaims the crucial value of wilderness for one who wishes to live freely. "The earth," he writes, "is not a mere fragment of dead history . . . but living poetry . . . not a fossil earth but a living earth . . . the institutions upon it, are plastic like clay in the hands of the potter" (308–09). Moving toward conclusion, he turns to an explicit discussion of the value of wilderness. Part of his argument rests upon the notion that the wilds are not just for humans and that human institutions are fragile, tenuous, and not nearly as important as some of us want to believe they are: "We need to witness our own limits transgressed, and some life pasturing freely where we never wander. We are cheered when we observe the vulture feeding on the carrion which disgusts and disheartens us and deriving health and strength from the repast . . . tender organizations can be so serenely squashed out of existence like pulp . . . The impression made on a wise man is that of universal innocence" (318). Wilderness, Thoreau shows, has the wholly unique ability to make us broaden our conceptions about our place in universe. As Bradley Dean writes, "Conceptions developed within exclusively civilized frameworks generally and perhaps invariably reflect the insular, anthropocentric origin by falsely regarding humans as superior to or otherwise separate from nature." Wilderness restores balance, helps us realize our limitations and the fact that we are not the measure of all things (83).

Like many American writers discussed in this chapter, Thoreau admired deeply works of classic European literature, but the continent's ruins function largely as a metaphor for civilized, unwild (and hence inferior) Europe over and against the relative wildness of America. It is not old Europe but primitive wilderness that holds the keys to permanence and authenticity. In his essay "Walking," he writes of having gone to observe panoramas of the Rheine and then of the Mississippi River. Watching the former, he feels enchanted, as if he "had been transported to a heroic age, and breathed an atmosphere of chivalry." But he comes to realize that "*this was the Heroic Age itself*, though we know it not, for the hero is commonly the simplest and obscurest of men" (202), and it is with this realization that he proceeds with the next paragraph into the most famous passage from the essay: "The West of which I speak is but another name for the Wild; and what I have been preparing to say is, that in Wildness is the preservation of the world" (202). Whereas the civilized nations of Europe (he names Greece, Rome, and England) are "sustained by the primitive forests which anciently rotted where they stand" (206), the New World has the precious gift of wildness, the raw material for physical and spiritual freedom.

The Skeptics (Hawthorne and Melville)

Other major American writers contemporaneous with Emerson did not address the changing natural world in apocalyptic terms, though there are many other types of apocalypticism in their writings. Hawthorne, Melville, and Poe (discussed in Chap. 7) look inward, doubting where Emerson, Thoreau, and Whitman largely affirm the discovery of external nature as central to self-realization. But they are, to say the least, complex writers. Nature is precious and a sort of standard for these writers as it is for the Transcendentalists. Hawthorne's love for wild nature is apparent in his heavily allegorical works. Stories such as "Rappaccini's Daughter" belie a critic of scientism, and Hawthorne, who got to know Emerson and Thoreau while living in Concord in the 1840s and 1850s, found a solace in nature unavailable in the city. His Preface to *Mosses from and Old Manse* is autobiographical and almost Thoreauvian in its appreciation of wild nature and the kindredness of humans to it. (It names and draws from the thought of both Thoreau and Emerson.)

Yet external nature itself is not an explicit focus in his fiction, which, early and late, revolves around the inherent evil (and goodness) of humanity. Known for a period as the American Byron, Hawthorne's experiences in Rome, the basis for *The Marble Faun*, appear to have changed his thinking about nature. Says Kenyon, one of the novel's main characters, "We all of us, as we grow older . . . lose somewhat of our proximity to Nature. It is the price we pay for experience" (195). Brenda Wineapple writes, regarding Hawthorne's narrator in this, his last completed novel, that the writer pieces together the best he can those things "slated for extinction . . . Writing as restoration, memorialization, and a defense against ruin: these are doomed to failure. And that, finally, is the lesson of Rome" (327).

Any rejection of anthropocentrism is difficult to pinpoint in a body of fiction so thoroughly devoted to his recurrent literary dissections of the human heart. In his apocalyptic-titled story "Earth's Holocaust," for example, written in the 1840s during a period of reform, mobs of reformers build a bonfire to burn weapons, liquor, books, and then more abstract "ideas" such as religious abuse, tyranny, and so on. The rub of the story is that a utopia is impossible because the human heart has not been thrown into the fire. Says the Satanic "dark-visaged stranger" near the story's close, "it will be the old world yet!" (159).[5] In "The Custom-House" he compares human ideas and institutions with ruins, Puritanism itself being a ruined building. Yet his brief story "Wakefield" indicates that we may very easily realize or undertake our own annihilation by simply moving from home to the next street for twenty years. No one will miss us all that much and not for long. "The Ambitious Guest" is based on an actual 1826 incident in which a rock slide in the White Mountains of New Hampshire buried an entire family and their inn along with, perhaps, an inn guest. Given the catastrophic rockslide at the end of the story, the dialogue regarding grand ambitions is filled with dramatic irony. The family of the inn holds humble aspirations in life, while their nameless young guest desires fame, a long-lasting legacy, which was not to be: "His name and person utterly unknown; his history, his way of life, his plans, a mystery never to be solved; his death and his existence, equally a doubt!" (88).[6]

Melville's views were similar in ways to those of his older friend Hawthorne in a rejection of Transcendentalism or nature as a necessarily benevolent concept. Perhaps his most important statement about this is found in the most celebrated section of his greatest work, "The Whiteness of Whale" in *Moby Dick*, in which Ishmael attempts to explain Ahab's inscrutable hatred for the white whale, which has the "pallid hue" of the dead, though it is alive (162). Similarly, the city of Lima, says Ishmael, is still white from the aftereffects of the great earthquake of 1746 that destroyed the city in a manner of minutes; "this whiteness keeps her ruins for ever new; admits not the cheerful greenness of complete decay; spreads over her broken ramparts the rigid pallor of an apoplexy that fixes its own distortions" (163). Melville's grim picture of white death doesn't allow for the romantic image in countless stories and poems in which ruins are overtaken by living green foliage. Melville positively rejects the Emersonian-Thoreauvian view of nature as benevolent in one of the chapter's most discussed passages, as "all deified Nature absolutely paints like the harlot, whose allurements cover nothing but the charnel-house within" (165).

Echoing Byron, who felt himself "a ruin amidst ruins," for Melville, whiteness is an image of the incipient ruins within us all. Melville was much less autobiographical in his writing than Byron, but one wonders how much his recent commercial flops as a novelist contribute to such a dark vision of the world. (*Moby Dick* would be yet another commercial flop, and his subsequent novel, *Pierre, or the Ambiguities*, would nearly finish his career as a novelist.) Narratively, the white motif as well as the green one, concludes in the last long chapter in the novel, as Ahab and the *Pequod* crew have chased the whale for the third and final day. Preparing to follow the white whale again, Ahab bids farewell to the masthead and spies some green moss: "What's this?—the green? aye, tiny mosses in these warped cracks. No such green weather stains on Ahab's head! There's the difference now between man's old age and matter's" (421). Hence Ahab affirms a difference between vegetable and human nature: the former regenerates while the latter does not.

Melville wrote *Moby Dick*, an apocalyptic novel almost from its first page, at a time when the nation was obsessed with the millennium, and the novel makes ironic use of this context. Many of its prefatory Extracts, "Supplied by a Sub-Sub-Librarian," hint at the whale's destructive capabilities. The novel gains some of its considerable power from evoking biblical and historical apocalyptic references. Walking the gloomy streets of New Bedford in Chap. 2, Ishmael sees flying ashes and wonders, "are these from that destroyed city, Gomorrah?" (24). The early appearance of Elijah in the novel (Chap. 19) along with many other eschatological references foreshadow disaster for the *Pequod* at the novel's conclusion. The whaler *Jeroboam* is plagued with an epidemic (250), while another ship, the ironically titled *Rose-bud*, has tied a "blasted whale" to it that smells worse than a plague city (313). The entirety of "The Try-Works," with a crew resembling demons in hell, is filled with such imagery—though the chapter concludes with the apparently wry moral, "Give not thyself up, then, to fire, lest it invert thee, deaden thee; as for a time it did me" (328). In "The Fossil Whale," Ishmael directly places the narrative within an apocalyptic reference as he says, considering the literal and figurative massiveness of his subject, "Give me Vesuvius' crater for an inkstand! . . . No great and enduring volume can ever be written on the flea, though many there be who have tried it" (349). A few pages later, in "The Pacific," the sea seems to Ishmael a giant graveyard—a metaphor that recalls several Bryant poems in which the earth is a giant graveyard (but is also looks forward to Marianne Moore's "A Grave": "the sea has nothing to give but a well excavated grave" (5)). And on the last page of the novel, sharks and birds appear to conspire with the white whale against the *Pequod* crew. A sky-hawk, representative of the archangel Gabriel (whose appearances in the Bible carry momentous messages) shrieks as the *Pequod* sinks (426–27).

As the novel moves toward the long chase of the white whale and conclusion, Ahab and the shadowy harpooner Fedallah get into a whale boat, the underside of which sharks tap with their tails, creating a "sound like the moaning in squadrons over Asphaltites [Dead Sea] of unforgiven ghosts of Gomorrah." The two appear to Ishmael as "the last two men in a flooded world" (377). Signs of doom, as in the typhoon that hits the *Pequod* "like an exploding bomb upon a dazed and sleepy town" (379) and the electric fireballs ("corpusants") that scare the superstitious sailors (381), accelerate in the pages just before the final chase of the whale. Near the end, the whale's head, earlier designated as a battering ram (the title of Chap. 76), is, like a deity, "predestinating" (not predestined, as if the agent of God); "Retribution, swift vengeance, eternal malice were in his whole aspect" (425). The detail echoes Elijah in Chap. 19, as he says to Ishmael and Queequeg, "Ye've shipped, have ye? Names down on the papers? . . . Any how, it's all fixed and arranged a'ready" (88).

Where Emerson and Thoreau (and Cole) turn a critical eye on a long-developed Europe, Melville is more prone to subverting the exceptionality of the New World. A large portion of his first novel, *Typee*, explores this idea. Although (or perhaps because) the native islanders appear to practice cannibalism, their way of life is superior to that of Euro-Americans. In the short but packed chapter "Brit" in *Moby Dick*, he uses a polemical humor very similar to that of Thoreau (who was no believer in American exceptionalism outside of its abundant wilderness) to undermine American superiority, though the critique expands toward humanity: "Columbus sailed over numberless unknown worlds to discover his one superficial western one . . . however baby man may brag of his science and skill . . . man has lost that sense of the full awfulness of the sea which aboriginally belongs to it" (224). All the preternatural terrors exist despite human claims to civilization. The earth remains and will remain mostly covered by the sea, "an everlasting terra incognita" (224).

Melville's great protomodernist story "Bartleby, the Scrivener: A Story of Wall-Street" (1853) evokes the view not of a boundless, meta-human ocean but of a much more everyday white wall that the title character views in his "dead-wall reveries." One way to read the story is a critique on the rat race of capitalism, the national seat of which remains, as it was in the 1850s, Wall Street. Melville suggests incipient ruin for this world in his story. On Sundays, Wall Street, is as "deserted as Petra; and every night of every day it is an emptiness" (27), and it is here that Bartleby, alone in the law office, becomes an "innocent and transformed Marius brooding among the ruins of Carthage" (28).[7] Later, the narrator describes Bartleby as "A bit of wreck in the mid Atlantic" (32)—a phrase that might also be applied autobiographically to Melville, who had lost most of his audience by the 1850s. The phrase also connects Bartleby (and perhaps Melville) with Byron ("a ruin amidst ruins")—the story's enterprising narrator, in a seeming throwaway line, calls Byron "meddlesome" (20).

After Bartleby begins preferring not to fulfill his work duties, the lawyer narrator refuses to banish him: "I should have as soon thought of turning my pale plaster-of-paris bust of Cicero out of doors" (21). Since Cicero is known for his statesmanlike eloquence, the reference is ironic, since Bartleby, who says only "I would prefer not to" and a few variants, remains completely silent and passively resistant. (Later, Bartleby, refusing to engage in explanations for his idleness to the narrator, keeps "his glance fixed upon my bust of Cicero, which as I then sat, was directly behind me, some six inches above my head" (30).) As the narrator attempts to rid himself completely of Bartleby, giving him a generous amount of money, Bartleby "answered not a word; like the last column of some ruined temple, he remained standing mute and solitary in the middle of the otherwise deserted room" (33).[8] These metaphors suggest the impermanence of the capitalist system that undergirds the (ultimately) meaningless activities of its human machines as scriveners function as human photocopiers. A modernist, posthuman trope, humans-as-machines metaphors, reflecting industrial realities, accelerated in the later nineteenth century and peaked in the early twentieth. They gave way, increasingly, to tropes in which humans are displaced by mechanization, an idea that will likely dominate the twenty-first century and beyond.

Notes

1. Shepard Kreech III's *The Ecological Indian: Myth and History* (New York: Norton, 1999) demythologizes romanticized conceptions of Indians living in harmony with the land. In *Changes in the Land: Indians,*

Colonists, and the Ecology of New England (Hill and Wang, New York: 1983), William Cronon cites firsthand accounts of English colonists such as Roger Williams, William Wood, Thomas Morton, and Timothy Dwight of Indians regularly burning forests for hunting and agriculture (47–51).

2. A coda to Bryant: in 1995, archaeologists near the site of the pre-Columbian city Cahokia (near modern St. Louis) happened upon the remains of many houses around 1000 years old amid a construction site for what would be a modern housing subdivision (see Timothy R. Pauketat. *Cahokia: Ancient America's Great City on the Mississippi*. New York: Viking, 2009. 120).
3. I do not suggest that Irving and Cooper held identical political views, which were enmeshed in the contexts of early nineteenth-century America. Both were skeptical toward democracy, the "tyranny of opinion," but Irving was a conservative while, by the 1830s, Cooper became a radical, and, unlike Irving, developed contempt for the financial aristocracy. See Arthur M. Schlesinger, Jr. *The Age of Jackson*. Boston: Little, Brown, 1945: 238–239, 375–380.
4. Robert Hughes notes that, like Claude Lorrain, Cole never learned how to draw the human figure or face competently (141).
5. Hawthorne's views on reform are not so far from those of Emerson or Thoreau; cf. *Walden* regarding the need for one to first reform oneself by cultivating the brain, a "few cubic feet of flesh" (5).
6. The incident would, unsurprisingly, also be a subject for a Cole painting, *A View of the Mountain Pass Called the Notch of the White Mountains* (1839).
7. Gaius Marius (BCE 157–86) was a Roman general and consul. According to Plutarch, after being banished by governor Sextilius from Africa, Sextilius' official asked Marius if he had a message for the governor. Marius replied, "Tell him, then, that thou hast seen Caius Marius a fugitive, seated amid the ruins of Carthage" (*Lives Vol. IX*. Trans: Bernadoote Perrin. Cambridge: Loeb, 1920: 577).
8. Melville would visit Rome in 1857 and refer to the statues of Rome as "mute marbles" in his lecture "Statues in Rome," which ends with a quotation from Byron's *Childe Harold's Pilgrimage*: "when Rome falls, the world" (*Piazza* 409).

CHAPTER 6

Jeffers's Inheritors: "Transhuman Magnificence" in Late-Twentieth Century American Poetry

One tumultuous century after Thomas Cole's apocalyptic vision in *The Course of Empire* of an American civilization sinking due to excess, his vision resounded in a new way with Robinson Jeffers (1887–1962), whose poetry argues vociferously that humans hold an irrelevant place in the large scheme of the natural world and condemns humanity, insisting that it will burn itself out within a few generations. For Jeffers, Western civilization, America in particular, is at the point of (to use Cole's word) consummation, with destruction soon to follow. "Pain and terror," he writes in "Apology for Bad Dreams," are "not accidents but essential" (*Collected* 37). Possessed of an innate impulse to exploit and destroy the natural world, man is also destroying himself—"Now he's bred knives on nature turns them also inward" ("Science" 6). Hence, writing in the wake of World War II, "the P-38s and the Flying Fortresses are as natural as horseflies" ("Calm and Full the Ocean" 7); one disappointed with corruption may as well "Be angry at the sun for setting" ("Be Angry at the Sun" 5), and, at odds with human folly, "rock, sea and stars" are "fool-proof and permanent" ("The Old Stonemason" 20). The best and only hope for humanity lies in realizing its own anthropocentrism and acting ethically in accordance with this knowledge:

> We must uncenter our minds from ourselves;
> We must unhumanize our views a little, and become confident
> As the rock and ocean that we were made from. ("Carmel Point" 13–15)

B.L. Moore, *Ecological Literature and the Critique of Anthropocentrism*, DOI 10.1007/978-3-319-60738-2_6

Jeffers's widely anthologized poem "The Purse-Seine" dramatizes this apocalypse by a combination of narrative and commentary. Observing sardine fishermen in motorized boats off Monterey Bay, the speaker notes the beautiful and terrible scene in which the fish realize they are caught, as they "wildly beat from one wall to the other of their closing destiny" (9). Next, the speaker is situated atop a mountain at night and looking, not at a shoal of fish, but at the lights of a "wide city" (14–15). The purse-seine becomes a metaphor for the inevitability of the downfall of civilization: "We have geared the machines and locked all together into interdependence; we have built the great cities; now/There is no escape" (17–18). Humankind is incestuously over-dependent on itself and unable to see anything but itself as the measure of all things, thereby insuring its downfall. The circle of the nets is closing, and government can now only expedite the "inevitable mass-disasters" (20–23). For Jeffers, what many (including, presumably, the chief decision makers) deem as "Progress" (24) is in fact nothing less than a building toward mass disaster. Humility and serious poetry (Jeffers writes elsewhere) may provide humanity with a little hope, but apparently, even though the "inevitable mass-disasters/Will not come in our time nor in our children's" ("The Purse-Seine" 20–21), Jeffers sees no way of stopping civilization's downfall (see B. Moore 132–33).

As much as any twentieth-century poet, Jeffers sought to create poetry in step with modern science, particularly the long-established but only sporadically recognized fact that the earth and the humans that inhabit are not the center of the universe. Poems such as "The Great Explosion" and "Nova," among many others, posit that any philosophy that ignores or distorts science cannot be taken seriously (see Karman 90). It would be, in Nietzsche's phrase, "all too human" to dismiss Jeffers as a mere misanthrope; for Jeffers, even "poor doll humanity has a place under heaven," though humans are "born of the rock and the air, not of a woman" (2.418). Yet scientific evidence, for Jeffers, pronounces definitively that not only are humans not the center of the universe; we are, in our hubris, the lowliest of creatures. Even as we persist in our self-importance, it is only after we are gone that the earth will begin to heal (see "Summer Holiday" 202).

His poems with a pointed opposition to an anthropocentric view of the world range back to *Tamar* (1923), but they grew more pointed with *The Double Axe* (1948). His previous volume of poetry, *Be Angry at the Sun* (his celebrated translation of *Medea* appeared between the two volumes), expresses antianthropocentrism by contrasting man with animals; whereas, as he writes in "Faith," the latter are governed by instinct, "man needs lies," and through these, civilizations fall apart (3.3). The clear gist of the poem is the question of what is so great about man that he should consider himself the center of the universe. Unlike Greek and Roman antiquity, the twentieth century is characterized by "immense vulgarities of misapplied science and decaying Christianity," yet poetry is, he suggests, capable of restoring a small amount of clarity ("Prescription of Painful Ends" 3.14). Jeffers had, like many other modern writers, considered World War I a sham, the end of humanist and Judeo-Christian values (see Karman 41). A reaction to the war, in the few years following, Jeffers's poetry turned more fully toward wild nature but also tended toward the violent.

With *The Double Axe*, World War II much on his mind, Jeffers identifies himself with the prophet Cassandra, and he appropriates the Calvinist tenet (learned, in part, from his father, a Presbyterian minister and Old Testament scholar) that people are depraved, or, as he puts it in "Original Sin," "all are vicious" (3.204). The death of an orca is beautiful because humans are not involved in it; it is a creature possessed of "no lies, no smirk and no malice," elements which darken the earth ("Orca" 3.206). He posits in "Quia Absurdum" the "inherent nastiness of man and woman"; the stars themselves are "man-devouring" (3.213). An old man in the longer poem "The Inhumanist" underscores a pantheism of which humans are a part, but if he could elect one animal to rule the world, he'd "choose tiger or cobra but nothing cruel, or skunk/But nothing foul" like man (3.259). Man, he writes, "is no measure of anything" (3.264); he lives only to defile (3.260). A number of later poems have implications for ecofeminism, as the gender-neutral "man" increasingly gives way to the female, who has more in common with the earth, which is itself a living female (3.430). Echoing writers from Seneca to Diderot, Jeffers espouses a materialistic view, but the entire universe is alive and conscious, all part of one being, "God's life." Only man is "cruel and bloody-handed and quick-witted, having survived/Against all odds" (3.433). Jeffers's conception of God, then, is very close to the Spinozaon, pantheistic one.

Discussed a few years after the conclusion of World War II, Jeffers's central statement on humanity is realized in his tellingly titled concept of "inhumanism." Less a philosophy than a poetic mythology, Jeffers elaborates his viewpoint in the Preface to *The Double Axe* as

> a shifting of emphasis and significance from man to not-man; the rejection of human solipsism and recognition of the transhuman magnificence. It seems time that our race began to think as an adult does, rather than like an

> egocentric baby or insane person. This manner of thought and feeling is neither misanthropic nor pessimist, though two or three people have said so and may again. It involves no falsehoods, and is a means of maintaining sanity in slippery times; it has objective truth and human value. It offers a reasonable detachment as rule of conduct, instead of love, hate and envy. It neutralizes fanaticism and wild hopes; but it provides magnificence for the religious instinct, and satisfies our need to admire greatness and rejoice in beauty. (vii)

Inhumanism and the recognizing of transhuman magnificence are not the same ideas, though they are, perhaps, different sides of the same coin: a poetic-scientific understanding of the human context within the greater nature world.

Certainly, Jeffers was not the first writer of note to question or dismiss anthropocentrism. As previous chapters show, the idea is suggested by writers of classical antiquity (e.g., Empedocles, Lucretius, Seneca) and Eastern philosophies and religions. The rise of science, the rediscovery of literature from Greek and Roman antiquity, and modern science and the secularization of Western culture simultaneously advanced humanity and decentered humans from the cosmos. The questioning of the human *telos* was developed subtly but surely in a wide variety of contexts, mostly Christian, in poets such as Spenser and Donne, among others. It became dechristianized in the work of key Enlightenment thinkers such as Diderot and other materialists. It is an idea brooded upon by such Romantics as Wordsworth, Shelley, Byron, and, later, Tennyson, and it is applied in different ways in America by Thoreau, Muir, and a variety of writers into the twentieth century.

Robert Frost, in his traditional form, confronts the notion of the human place in a world not altogether supportable by traditional beliefs; this is to say that Frost was, like Jeffers, a modernist. (The poets came to admire one another's work.) More consciously poetic than Jeffers, Frost typically couches his often witty ideas in a landscape in which the speaker tests the ideas of humanizing wild nature or nature as the product of a beneficent deity, but he (again, typically) unravels this idea (e.g., "Birches" and "Design"). Jeffers extends and magnifies (some may argue, overplays) the subtler wisdom of Frost's "On the Need of Being Versed in Country Things," as in "Fire on the Hills," which states, "Beauty is not always lovely; the fire was beautiful, the terror/Of the deer was beautiful . . . The destruction that brings an eagle from heaven is better than mercy" (2.173). As poems such as "Once by the Pacific," "Fire and Ice," and "Desert

Places" indicate, Frost has much to say about hubris and hate as well as a gigantic universe of which humans are certainly not the center.

One of the more prominent inheritors of Jeffers's poetic vision is anthropologist and naturalist Loren Eiseley, who, as a young man, met Jeffers. Eiseley wrote poetry, but he is known mostly for his prose, which seeks an understanding for the common ancestry and interconnectedness of all living things, along with the deep implications proceeding from this understanding. In one of his most highly regarded books, *The Immense Journey*, he alludes to Jeffers's "Apology for Bad Dreams" in writing, "Men, troubled at last by the things they build, may toss in their sleep and dream bad dreams" (162). Eiseley concludes that we do not live with a proper understanding of ourselves and our fellow, kindred living beings and our place on the earth together within the context of geological history. Repeatedly, Eiseley writes of the lonely ability of humans to project themselves into other lives (46). We can see into the past, but we stop at the present, as if we've arrived, "the culmination and the end," as if "when we go, so will the universe" (57). Eiseley writes of having received letters concerning his "lack of faith in man" (24), but the idea that man began in a lowly position—the Paleocene rat is, he writes, the "father of mankind" (8) —is one to celebrate. We are coming to understand our place on the earth. The "wounded outcry of the human ego" was not dominant in the past (150), and, near the end of the book, he imagines, post man, the "most beautiful sight . . . birds taking over New York" (187).

Another important prose inheritor is Edward Abbey, who appropriates Jeffers's "Hurt Hawks" in *Desert Solitaire*: in coming to understand the symbiosis of the snakes living near his trailer at Arches National Monument (now National Park), he states, perhaps hyperbolically, "I'm a humanist; I'd rather kill a man than a snake" (20).[1] Here as elsewhere, Abbey appears to veer toward a misanthropy rooted in a deep doubt about human progress and industrialism, especially as it has emerged in his beloved American Southwest. More than once, Abbey echoes Jeffers in the optimism he holds based on the conviction that Western civilization is on the road to termination as a result of its own greed. Diane Wakoski shows that Abbey shares some of the fundamental views of Walt Whitman as well as Robinson Jeffers. Abbey, she posits, does not share Whitman's humanist choice "to believe that all mankind could be filled with love if it would, and that slavery, war, and other ignominies will be wiped away when his bigger vision is obtained." Nor does he quite go so far as to argue that man is a "evolutionary mistake," as Jeffers believes (169). The point at which Abbey converges with and synthesizes these visions is his belief that, for the most part, the human heart is good, but "in large numbers humankind is trouble" (174). I will have a bit more to say about Abbey in Chap. 8.

I turn to a discussion of how seven postwar American poets—Kenneth Rexroth, A.R. Ammons, Gary Snyder, Wendell Berry, Mary Oliver, Lucille Clifton, and Joy Harjo—contribute to the discussion of anthropocentrism in their work. My reference to them as Jeffers's "inheritors" has little to do with his direct influence on their poetry. Unlike Eiseley and Abbey, none of them appear to have acknowledged Jeffers as a major influence, at least two have directed criticism at him, and all of them are generally more hopeful toward humans than Jeffers. Insofar as "inhumanism" is an intellectual movement, it is one with a single member. Yet Jeffers's vision of a world beset by unwise human actions is arguably more pertinent now than it was during his lifetime. A vision of a world that is not human centered, whether we call it inhumanism, antianthopocentrism, or something close, underscores Jeffers's continued relevance. All the writers I discuss have in various ways confronted and subverted anthropocentrism and offered alternative, poetic visions that seek a more ethical and more science-based view of our place in the natural world. These poets have "inherited" Jeffers's most forceful argument and applied it in their own ways in different contexts.

My inclusion of Kenneth Rexroth as a Jeffers "inheritor" may seem curious given Rexroth's statement that "Jeffers's verse is shoddy and pretentious and the philosophizing is nothing but posturing" ("Poets" 215). Yet the two poets are aligned by their poetic celebration of California landscape, and like Jeffers, Rexroth is preoccupied with the relative longevity of rocks; some rocks, he writes, are "more durable/Than the configurations of heaven," while future species will see "stars in new clusters" (*Complete*, "A Lesson in Geography" 189). Like Jeffers and all the poets discussed in this chapter, Rexroth was discontent with the idea of progress in the twentieth century, "the Century of Horror," albeit in a more overtly politically leftist vein than Jeffers (Author's Note 233). Both poets are concerned centrally with the interrelationships between humans and wild nature, as Rexroth puts it, humans as "a microcosm in a macrocosm" (*In the Sierra* 119). We take these relationships for granted, or else we deny them altogether. His poem "Toward an Organic Philosophy" records the poet's thoughts on a wholly nonhuman Sierra landscape; only at the end of the poem does he step back and consider the human context through quoting the nineteenth-century Irish scientist John Tyndall:

". . . the concerns of this little place
Are changed and fashioned by the obliquity of the earth's axis,
The chain of dependence which runs through creation,
And links the roll of a planet alike with the interests
Of marmots and men." (166–67)

Rexroth often relates the dark mysteries of the outer, practically infinite universe with the inner, momentary, and private. Such a poem is "Inversely, As the Square of Their Distances Apart," in which the speaker becomes conscious of the relationships between the greater regeneration of nature with human copulation (214–16). In his 1963 book *Literature and Science*, Aldous Huxley singles out Rexroth's "Lyell's Hypothesis Again" as one of the few poems to suggest an understanding of modern science (60). In the poem Rexroth employs Lyell's attempt to explain the earth's past by present causes through a sort of narrative built around sensuality, as the speaker and his companion (perhaps literally his wife) are "Naked in the warm April air . . . Over our heads"; as she arises, the speaker observes that redwood cones are imprinted onto her flesh, just as they leave their marks on rocks (279). Like other creatures, we are all potential fossils. In another poem written around the same time, only a constantly changing nature is permanent, while (in a truly Jeffersian phrase) "the human race sinks toward/Oblivion" ("Andrée Rexroth" 289).

The later poem "Hapax" serves as a sort of ecological summing up amidst awareness of an ever-changing local environment set in an ineffably immense universe. The word "hapax" suggests a contingent, momentary set of thoughts or insights that change in the poet's perceptions. These thoughts reflect a local and universal mutability, as well as the poet's location in their midst: it is a wonder that the universe is so huge, but it is also a wonder that the human mind is capable of being conscious of this fact. Unlike Jeffers, Rexroth expresses some of the mysteries of existence in this cosmic backdrop without judging the humanity as petty. He thus creates a sort of bridge between the declamatory verse of Jeffers and the observational verse of A.R. Ammons.

Ammons's poetry is similar to Jeffers's in its fascination with and praise of nonhuman nature, and it also compares with Jeffers's in its disavowal of the centrality of humans in nature. Like Wallace Stevens, Ammons evokes the romantic notion that humans are the conscious animals, and like Emerson and Whitman, as well as Goethe, Ammons is obsessed with the unity of nature, the organic whole of which humans are only a part. He reflects this idea compactly in the title of his poem "One: Many." We are privy to only a sense of partial knowledge of nature's dynamism; seemingly insignificant events are crucial to understanding the "inevitable balances events will take" (*Selected* 40). Ammons steers away from what G.E. Moore termed the naturalistic fallacy, the confusing of the good with natural properties (conflating "is" with "ought"), traces of which emerge in some of Jeffers's poetry. Ammons's verse is, like Jeffers's, centered in a long, geological view of history. While virtually all poets seek to widen perceptions of the self and one's relation to the greater whole, the focus for both Jeffers and Ammons is a long view that precludes anthropocentrism. As Ammons states in compact form in "Gravelly Run": "it is not so much to know the self/as to know it as it is known/by the galaxy and cedar cone" (11).

Unlike Jeffers, who would place humans at the bottom of the chain of creation, Ammons posits a more objective notion that all of the parts are more or less equal. "You cannot come to unity and remain material," he writes in "Guide" (23). Humans are not negative beings, but neither are we more than a part of the gigantic whole; the earth, he writes, will be no lighter when he is gone, but "no heavier/with me here" ("The Account" 75). In "Still," he resolves to "find what is lowly" so he can establish his own roots of identity and measure his own significance (41). Digging into the dirt, he finds that "there is nothing lowly in the universe"—everything, "moss, beggar, weed, tick, pine, self, magnificent with being!" (42). Similarly, in his book-length poem *Sphere*, he writes that "if there are no boundaries that hold firm, everything can be ground into everything else" (61). Although such observations are grounded in modern science, it is an idea established much earlier, at least as early as Seneca, who in *Natural Questions* writes of the interconnectedness of nature, or elements moving in and out of one another, all things being a part of one great order, a position central to Stoicism. We can learn about the importance of things and interrelationships by looking outside ourselves. The Seneca of *Natural Questions* is positioned somewhere between Ammons and Jeffers in his understanding of interconnectedness, his view of unity and the "pinprick" of our existence (69, 137, etc.), but also in his tendency toward ecocatastrophe. The world can undo humanity very quickly: "A single day will bury the human race," he writes (51).

There are many more humans in Jeffers's poetry than Ammons's, which might seem at first glance odd given the former's "inhumanism." The two poets share a similar foundational, scientific view, and they both celebrate the wonders of the universe, but they reach different conclusions about the place of humans: Jeffers cites the contrasts where Ammons finds unity, all the while acknowledging (and perhaps celebrating), like Stevens, our limited perceptibility. As it is for Spenser's Mutabilitie, the world is governed by change, and since we and our language change as well, his poetry is the means of both underscoring that imperceptibility and seeking to account for it, as when he writes of the land as "a slow ocean"—all changes almost imperceptible for humans, "intermediates of stone and air" ("Delaware Water Gap" 94).

Like that of Jeffers and Rexroth, Gary Snyder's poetry is connected deeply to the land of California, and like Rexroth (who was an important influence on Snyder and other San Francisco Bay poets), Snyder has expressed reservations about Jeffers, namely his

> tall cold view . . . why did he say it
> as though he alone
> stood above our delusions. (*No Nature* 371)

Like most of the poets considered here, Snyder's interrogation of anthropocentrism is resolved, largely, in the idea that humans are not above (or below) the rest of the natural world but a part of it. In a letter he wrote to Edward Abbey, Snyder takes issue with Abbey's statement that India and the Far East are "the most miserable, most abused, most man-centered cultures on earth" ("A Letter" 182). His poetry, early and late, underscores his opposition to a hard anthropocentric view. In his first volume of poems, *Riprap*, published in Japan in 1959, the poet, positioned in a natural setting, writes that "All the junk that goes with being human/Drops away . . ." (*No Nature* 6).

The direct, skeptical questioning of the place of humans in the world is not a major theme in much of his poetry, but beginning with perhaps his most celebrated poetry collection, *Turtle Island* (1974), Snyder's antianthropocentrism intensifies considerably. To the question, "IS man most precious of all things?" he does not answer affirmatively but rather with a charge: "—then let us love him, and his brothers, all those/Fading living beings." White people (presumably) are "invaders" of Turtle Island (indigenous peoples' name for North America (xi)), and they "wage war around the world," but then he redefines the word "people" by including "Tree People," "Bird People," "Sea People," and "Four-legged, two legged, people," humans being one of the last-named group (*Turtle* p. 48). "Toward Climax" is a similar, more directly Jeffersian indictment of the effects of humans on their environment. At first, the world is pristine and innocent, but the poem proceeds with the growth of knowledge and civilization and climaxes with the American war in Vietnam and the acceleration of forest clear-cutting (*Turtle* 82–85). Although bitter about America's imperialism and its war on nature, he concludes that this is a relatively late defect: "These cliffs and the stars/Belong to the same universe," while "The little air in between/Belongs to the twentieth century and its wars" (*No Nature* 279). For Snyder, *Homo sapiens* in a more primitive state was at home in nature, whereas for Jeffers the species was murderous and ill-fitting from the start (cf. Jeffers's poem "Original Sin"). Notably, both Jeffers and Snyder reflect on the smallness of humans (and extolling transhuman magnificence) within the context of what they see as foolish, imperialistic wars.

Yet Snyder veers at times toward suggesting the insignificance not only of the human species but all other species as well as the earth itself. A later short poem, "For Nothing," imagines earth as a flower "hanging over vast/solid spaces" and ultimately "forgotten as all falls away" (*No Nature* 229). John Elder reads the poem as a sort of elegy depicting "the smallness and isolation of our blue-green, cloud-swirled orb, when viewed from orbit as in the famous NASA photograph" (710). It may also be read as a variation on a theme in Cicero's *The Dream of Scipio*, in which the shade of the famous Roman general Scipio Africanus notes the smallness of the earth from the vantage point of the Milky Way (*Republic* 6.16). Readers are left to conclude whether such poems are pessimistic or realistic. Raw data may lend credence to the former, but Snyder's Zen Buddhism suggests the latter. American poet Pattiann Rogers addresses this complex idea thus: "we have figures to prove [the tiny span] of our insignificance. Our lives are fleeting compared to the age of the earth, the history of the sun, the solar system, the Milky Way, a pulsar, a quasar" (4). Snyder, like other poets discussed in this study, gives "us poems that say what it feels like to accept the geological record" (Rhodes 240). Transhumanism is not an inherently pessimistic viewpoint but rather a scientifically based reality.

Snyder's *Mountains and Rivers Without End* (1996) is part Beat poetry (its earliest poems date from the mid-1950s), part travelogue, and part rumination on the human place in the natural world. "Night Highway Ninety-Nine" recounts travel in the Western United States, more a celebration than critique of the people he encounters; near the end of the poem, however, the speaker has grown tired of car exhaust and the highway, upon which the "squat earth-movers" resemble "yellow bugs" (302). He prefers instead to speak for hawks. The volume's final poem, "Finding the Space in the Heart," appropriates indigenous American myth and Asian religious thought with science (ecology, geology, anthropology) to suggest the human place in the natural world. Unlike Jeffers, Snyder finds a place for humans, but, as Coyote tells cottontail boy, you must "learn your place" (121).

Wendell Berry's Christianity would seem to place him in direct conflict with Jeffers's central position of a pantheistic God that bestows no special status for humans.[2] But many of Wendell Berry's poems evoke a natural world independent of humans, a species that is often found wanting in comparison to a simple, innocent natural world, and his faith directs him to see God's nature as sacred yet too often ignored and exploited by his fellow believers: since God loves the world, "how might a person offaith be excused for not loving it or justified in destroying it?" ("God and Country" 98). Echoing Jeffers's poem "The Inhumanist," Berry's "The Morning News" marks the violence in humanity and notes, "The serpent [is] gentle, compared to man" (124). Elsewhere, Berry indirectly channels Walt Whitman in Chant 32 of *Song of Myself*, which praises animals for being "self-contained" and not possessed of the uniquely human "mania of owning things." Images employing what might be called inhumanistic ideas appear in Berry's early poems. In "Observance," a river god hears townsmen singing to him "briefly as reeds/grown up by the water" (6), and "A Man Walking and Singing" begins with the lines, "It is no longer necessary to sleep/in order to dream of our destruction" (11). But the idea abounds in Berry's 1968 volume of poems *Openings*, which is rife with criticism of the Vietnam War and the human race in general against the backdrop of the wild nature around the writer's home in Port Royal, Kentucky. "The Dream" may be read as a sort of anthropocentric parable, though it also resembles, in short form, the passing of men from the noble golden to the toilsome iron ages in Hesiod and Ovid (cf. Snyder's "Toward Climax"). The speaker dreams of restoring nature to its pristine condition, so he removes bridges, roads, and fences and the like, but the old human urge arises, and he is suddenly "eager to own the earth and to own men. I find in my mouth a bitter taste of money," and he sees that "we have ruined in order to have" (64–65). In "The Sycamore," the names on the grave stones are fading in a cemetery that is growing wild, "clear to the rabbits and the wren" (66). "Dark with Power" terms humans "the invaders of our lands," who cause desertification and scarred hills (67).

In his long poem "Windows," the county is haunted by the ghosts of the old forest, which begin to rise in a postapocalyptic city; machines having devoured humans, "then/there will be a second coming/of the trees . . ." The speaker of the poem at times "thinks the earth/might be better without humans" (83). Berry veers here very closely not only to Jeffers but to Berry's friend Edward Abbey, who, as noted, has expressed similar ideas throughout his prose work. A dozen years later, inhumanism again appears in his poem "The Slip" (1980):

> The river takes the land, and leaves nothing.
> Where the great slip gave way in the bank
> and an acre disappeared, all human plans
> dissolve. An awful clarification occurs
> where a place was. Its memory breaks
> from what is known now, begins to drift.
> . . .
> Human wrong is in the cause, human
> ruin in the effect—but no matter;
> all will be lost . . . (1–6, 10–11)

Unlike Jeffers, Berry asserts the grace of a beneficent God who oversees justice in its true, nonhuman perspective, by way of "the clear eye/of Heaven" (15–16).

Several of Berry's "Sabbath Poems" similarly evoke the transience of human operations and the (often unwise) exploitation of the natural environment. In the practice of small farming, he writes, "the world is used/But not destroyed ..." (14), but his landscapes point regularly to the garish, the unwise, and merely monetarily profitable. Echoing Jeffers, though in a Christian framework, he asks how humans might "pray to escape the catastrophe/that we have not the vision to oppose and have/therefore deserved ..." (110), yet he notes in another Sabbath Poem that Nature's healing "will come in spite of us, after us,/over the graves of its wasters ..." (47). Although his poetry holds the possibility of grace and for (the title of one of his most anthologized poems) "The Peace of Wild Things," Berry's pessimism toward human progress resembles Jeffers's. A later Sabbath Poem (208–09) describes a ruined world concocted first by planners in offices and then realized by bulldozers, clear-cut forests, poisoned rivers, and a deluded citizenry. It is unclear if the world of the poem is in the present or future. Outside the house at night, looking at "Heaven's lights" makes one aware of "our smallness," yet the lights from distant cities suggest (echoing Jeffers's "The Purse-Seine") human certainty about progress, "as if we will have no light/but our own and thus make illusory/all the light we have" (*A Timbered* 198).

Like Berry's, Mary Oliver's poetry reflects a spiritual search for meaning to human existence in a natural environment. A heron she observes is a "blue preacher" (p. 81), and the moss could give a "lecture" on "spiritual patience" (129). All of Oliver's nature poetry (and a high percentage of her work is nature poetry) is infused with a sense of temporal, ever-changing humans in a less temporal nonhuman landscape. Since Oliver persistently seeks a reconciliation of human with nonhuman (a tradition established in America by Anne Bradstreet in "Contemplations"), traces of Jeffers are, perhaps, remote. Yet the focus on the temporal human looking at and learning from an equally (or more) transient wildlife in the more permanent backdrop of wild nature marks her poetry as "transhuman," though it is certainly not "inhumanist."

Maxine Kumin, in a review of an Oliver collection, is struck by "the exactitude of [Oliver's] imagery, by her daring marriages of animal, vegetable and mineral kingdoms to the human condition, and by her slightly amended transcendentalism, which seems to allow for a stoical embrace of her own mortality" (19). In an early poem about the burial of a dog, Oliver notes that birds, cranes, and dogs are not presumptuous and self-centered, unlike humans, who are aware of their own mortality (15). Like our "dumb wild blind cousins," we lose consciousness, and this binds us with nonhuman nature (57). Hoofed and flying animals, she claims, "do not cry or argue" (214)—echoing Berry and Whitman's animals in *Song of Myself* Chant 32. Oliver's empathy for wild nature extends beyond animals; echoing Wordsworth's "I Wandered Lonely as a Cloud," in "White Flowers" the speaker lies in a field of flowers and senses "the porous line" between her own body and those of the flowers (59). Such plant/animal/human convergences are common in her work, but these poems are also, like many by Wallace Stevens, about the unknowability of nature. The landscapes themselves—along with their natural flora and fauna—constitute the bulk of Oliver's verse, but, more frequently than Ammons, Oliver steps aside to comment directly on the landscapes. She observes a black bear and comments: "there is only one question: how to love this world" (70), then she returns to observing the bear. Wild geese show us our place "in the family of things" (110). Our transient lives bind us with other living things; she notes that the "terror is that nothing/laments the narrow span" (214). Though her viewpoint is finally not Jeffersian, the implications of her self-conscious exploration of a proper (scientific and ethical) balance between human temporality and our relative insignificance, like Jeffers, suggests a much humbler human presence on earth and an interrogation of the idea of progress.

Much less known as an ecopoet than Oliver, Lucille Clifton addresses and dismantles class, racial, and gendered conceptions that humans possess differing degrees of inherent worth, and she comments particularly on the white male degradation of black female bodies.[3] Synchronously, her poetry, early and late, however, also links the exploitation and degradation of bodies with the exploitation and degradation of the planet. In one of her most commonly anthologized poems, the speaker wishes upon her sons cramps, a surprise menstruation, and other experiences unique to women as a cautionary against "arrogance in the universe" ("wishes for sons" p. 382). Over forty-five years as a published poet, Clifton was a vital, consistent supplier of antidotes to arrogance with regard to race, gender, nationality, and the understanding of our place in the greater world. Clifton is a humanist and, though sometimes indignant, is also optimistic, seeking reconciliations; her humanity embraces and explores both consciousness of our present moment on earth as well as an understanding of our species' relative smallness. Although her poetry is most often set in urban areas, she also seeks to account for the natural world and explore our place in it.

Clifton's ecological sense extends through the personal, the political, and the socially-minded, and it extends to a disavowal of a hard anthropocentrism. Life is a gift, but it is also brief and does not validate arrogance toward nature, human and nonhuman. People soon die and become the "bottoms of trees," so we should "bear responsibility to something/besides people" (56), she wrote in her first collection of poems, *Good Times* (1969). Her poem on an Old Testament frame "what the grass knew" suggests that the grass was made not just for Adam, "that what is built/is finally built for others" (350). In an untitled late poem, humans are not the wisest or blessed, are not the sole or "favorite servant/of ones lord" (621). And in the still later, apocalyptic "godspeak: out of paradise," humans are "mad children" who, placed in safe beds, have responded by tearing the crib apart and now sit among ruins (690).

Her poem "the mississippi river empties into the gulf" expands the theme of the earth as a giant (female) body and also underscores the arrogance of assuming that the present is the only reality that matters. As one looks at the river, a Heraclitean symbol, one is led not to the conclusion that it exists for the present but to the reality that we are placed on earth at a brief point in time. Always seeking an understanding of her environment, Clifton also reflects in the poem a new landscape during a semester teaching at Memphis State University (now University of Memphis) in 1994. The river is a part of the circulation system of a living being. The poem concludes by invoking the idea that we constitute but a brief point in time, as someone stands on the river's edge, "whispering mistakenly:/only here, only now" (505). Perhaps only tangentially Jeffersian in her worldview, such a poem bears a striking resemblance to Jeffers's typical interrogation of human presumptuousness to a supreme universal position, as in "Wise Men in Their Bad Hours," in which a grasshopper says, "What does the future matter? We shall be dead" (1.10).

The poetic theme of inhumanisim by nature holds strong implications not only for ecology but apocalyptic themes because it often envisions a world without or with fewer humans. Where Clifton veers toward the apocalyptic within a female African American historical context, many American Indian writers do so by challenging white hegemony and showing the historical dangers and tragic results of not only a species but a race within the species that has repeatedly wielded authority over the rest of the natural world. As I note in the first chapter, Linda Hogan connects the mistreatment of American Indians, who endured their own holocaust well before the one enacted in World War II, with the war that extends to the natural world, and this leads her to conclude that "what happens to people and what happens to the land is the same thing" (89). One poet who explores this idea is Joy Harjo (Muscogee), who, like other American Indian poets writes much about the hereditary of linking humans with nonhumans as well as the questioning of human predominance, especially through the context of white imperialism. She urges readers to "Remember the earth whose skin you are . . . we are earth . . . Remember the plants, trees, animal life who all have their/tribes, their families, their histories too" ("Remember" p. 42).

As with other Amerindian poets, she calls upon us to listen to nature—animals, plants, rocks—because it is speaking to us. In "A Map to the Next World," she writes, "In the legend are instructions on the language of the land," though we have forgotten this gift (129). The notion that nature speaks to us is of course not solely a *topos* for Amerindians. Annie Dillard, for example, intertextually qualifies and elucidates this idea in her own way in her much-published essay "Living Like Weasels": "I don't think I can learn from a wild animal how to live in particular—shall I suck warm blood, hold my tail high, walk with my footprints precisely over the prints of my hands?—but I might learn something of mindlessness, something of the purity of living in the physical senses and the dignity of living without bias or motive" (15). Yet the historical basis for a poet such as Harjo gives such ideas a powerful rhetorical force.

In her prose poem "Transformations"—echoing Clifton's "what the grass knew"—an ambulance comes "to rescue an old man who is slowly losing his life. Not many can see that he is already becoming the backyard tree he has tended for years, before he moves on" (84). The chronological progression of Harjo's *How We Became Human: New and Selected Poems: 1975–2001* reveals an emerging, recurring theme regarding our kinship with the stuff of the universe, the earth and the stars. She writes, "Our souls imitate lights in the Milky Way. We've always known where to go to become ourselves again in the human comedy. It's the how that baffles" (114). The title of the collection suggests, following both Indian myth and science, that humans are based on the earth, are a part of it. The centrality of water in creating life and in, eventually, bringing about human consciousness—how we became human—is the subject of "It's Raining in Honolulu," her city of residence. Although humans are a part of nature, "A human mind is small when thinking/of small things" (137).

In "Hold Up," the speaker is posited between transcendence and the unpleasant realities of everyday life. Her mother has told her that her father is the sun, which created "a nimbic web that embraced us"; people, a part of the wonders of the universe, die "without knowing they breathed planets." But a hold up robs her of money and of her idealism. Amidst mixed signs, she states, perhaps provisionally (though twice), that "*Humans were created by mistake*" (146–48). In "Morning Prayers," she recalls how, when younger, she looked to the Sangre de Cristos mountains, with "a notion of the sacred." She had believed in a saving vision that would carry everyone "to the top of the mountain/during the flood/of human destruction," but grown older, she states, "I know nothing anymore" (186–87). However, Harjo is more prone to praising humanity as essentially part and parcel of nature but decrying white pretensions to ownership: "this earth cannot be owned," she writes (156), and white men fool themselves and "the sleeping ones into thinking they've bought the world" (170).

In his directness and in his scientific viewpoint, Jeffers remains a sort of standard for the poetic interrogation of anthropocentrism, even though relatively few writers share fully his dim view of humanity. All seven of the poets discussed here temper his extreme inhumanism with a perhaps more objective understanding that humans are not above the rest of nature, but neither are we below it. This idea applies to many other postwar poets of note that space has compelled me to leave out, including Denise Levertov, W.S. Merwin, Robert Hass, and Pattiann Rogers. Perhaps few people in general would accept Jeffers's view of humanity, but many, perhaps most, people would accept (if not act upon) the notion that our species has ravaged the planet and that a humbler view of ourselves is in order. The questioning of a hard anthropocentrism is central in this realization, but to assume that such questioning is the same thing as wishing for the immolation of *Homo sapiens* is an extreme leap, even for pure Jeffersians. The questioning of absolute human supremacy is tempered by Snyder: "The/awareness of emptiness/brings forth a heart of compassion!" (*Mountains* 149). Ammons delivers a similar thought in more comically frank language: "where but in the very asshole of comedown is/redemption" (*Garbage* 21). If not Jeffers's pure inhumanism, then a kindred understanding of our humbler membership in the natural world along with a greater awareness of transhuman magnificence is necessary if not sufficient in addressing our present environmental crisis.

Notes

1. Abbey utilizes the second section of one of Jeffers's most famous poems, "Hurt Hawks," which begins with the line, "I'd sooner, except the penalties, kill a man than a hawk" (18).
2. Berry was attacked by deep ecologists in the 1980s for what they saw as a "shallow" ecology inherent in Christianity. See Chap. 7 of Bill Devall and George Sessions. *Deep Ecology: Living as If Nature Mattered.* Layton: Gibbs Smith, 1985. Berry addresses deep ecology in "Amplications: Preserving Wildness," *Wilderness* 50 (Spring 1987): 39–40, 50–54. In his review of Berry's *Home Economics*, Edward Abbey writes that Berry is "the best serious essayist at work in the United States," but he levels a charge echoing that of the deep ecologists: "The trouble with the concept of 'stewardship' is that the stewards tend to think they have the God-given right to exercise domination over the entire planet." Berry responded to Abbey, stating he sees "no inconsistency between this idea of stewardship and the idea of wilderness preservation" (http://www.canyoncountryzephyr.com/newzephyr/october-november 2010/pdfs/oct10-12-13.pdf).
3. Clifton's rhetoric is comparable with that of Delores S. Williams, who writes, "the assault upon the natural environment today is but an extension of the assault upon black women's bodies in the nineteenth century." Williams is a developer of Womanist Theology, which has "labelled this assault upon the environment and upon black women's bodies as sin," or more particularly "defilement" ("Sin, Nature, and Black Women's Bodies." *Ecofeminism and the Sacred.* Ed. Carol J. Adams. New York: Continuum, 1993: 25).

CHAPTER 7

Antianthropocentrism and Science Fiction Part I: From Antiquity to World War II

To a degree no less than that of writers discussed so far, including Jeffers, many works of science fiction (hereafter "sf") form a rich, varied, and philosophically interesting tradition of questioning anthropocentrism and some of the time even suggesting or explicitly espousing ecocentrism. Contemporary sf writers such as Kim Stanley Robinson and Joan Slonczewski exhibit subtle, sophisticated treatments of ecology in their widely-celebrated novels. By contrast, early sf writers have been noted for their obliviousness toward ecology. For example, Brian Stableford writes that "there are very few early stories with ecological themes." Early sf writers "were often oblivious to the simplest matters of ecology" in writing about life on other worlds. Among pulp writers, only Stanley G. Weinbaum "showed anything more than a rudimentary consciousness of the subject" (365). This chapter seeks less to overturn this judgment than to moderate it; although often relatively lacking in sophistication and superficial in its science, many sf works written before World War II contain ideas of continuing ecological relevance. Many early sf pieces dramatize fantastic situations in which antianthropocentric ideas emerge that anticipate Loren Eiseley's idea that "we see, perhaps inevitably, through human eyes alone. We see ourselves as the culmination and the end, and if we do indeed consider our passing, we think that sunlight will go with us and the earth be dark" (57).

It is no coincidence that, in the wake of the modern environmental movement—based on the work of the English Romantics and the New England Transcendentalists, grounded scientifically in the writings of

B.L. Moore, *Ecological Literature and the Critique of Anthropocentrism*, DOI 10.1007/978-3-319-60738-2_7

Charles Darwin, Aldo Leopold, and Rachel Carson, and arriving fully near the end of the 1960s—many sf writers would adopt environmental themes in their work. The subgenre of sf ecological catastrophe ("ecocatastrophe") literature flourished around this time and peaked in the early 1970s, in the wake of several widely reported global environmental disasters. Many ecological-minded sf novels, stories, and anthologies appeared beginning in the early seventies. The audience for such literature was and is still forced to think freshly about our species' place on earth. Science fiction, almost from its beginning, has been a popular genre, from pulp to cheap paperback, Hollywood, television, and the Internet. It has made inroads to academia within the last generation or so, and it remains a popular form of literature.

Although early sf remains largely ignored among ecocritics, it is, along with other popular genres of literature and pop culture, one of the small but underappreciated elements that have contributed to public appreciation of and wish to protect the environment. Perhaps it is merely coincidental that American sf started to flourish around the time John Muir began publishing his influential nature books at the turn of the previous century. The genre, its Golden Age, reached a highpoint around the time of Aldo Leopold's *A Sand County Almanac* in 1949, the publication of which led to the rise of the environmental movement in the United States, and with it the 1964 Wilderness Act, the founding of the Environmental Protection Agency and the establishment of Earth Day in 1970, and the Endangered Species Act of 1973.

The critique of anthropocentrism in sf is one manifestation of Brian W. Aldiss's famous definition of the genre—"*Hubris clobbered by nemesis*" (*Trillion* 26). In many stories, characters find themselves in situations that subvert the assumption that the human species is the only one that matters. I do not suggest that sf is an inherently antianthropocentric genre, but several sf writers have addressed its unique ability to account for the nonhuman. C.S. Lewis notes a "sub-species" of sf he terms "the Eschatological," which includes novels such as H.G. Wells's *The Time Machine*, Olaf Stapledon's *Last and First Men*, and Arthur C. Clarke's *Childhood's End*. Such works give "an imaginative vehicle to speculations about the ultimate destiny of our species." For Lewis, "It is sobering and cathartic to remember, now and then, our collective smallness, our apparent isolation, the apparent indifference of nature, the slow biological, geological, and astronomical processes which may, in the long run, make many of our hopes (possibly some of our fears) ridiculous" (65–66). Ursula K. Le Guin writes that sf "is almost the only kind of story that ever really admits of a world not dominated by human beings (or gods, animals, or aliens who act just like human beings)" (Introduction xv). And Frederik Pohl writes, "One of the things that most attract me to science fiction is its capacity to give what [American astronomer] Harlow Shapley calls 'the view from a distant star,' the perspective on our humanity through the eyes of nonhumans" (100).

In a 1933 preface to a collection of his own scientific romances, Wells writes of the genre's uniqueness in its potential for "looking at human feelings and human ways, from [a] new angle" (qtd. in Evans 178). As Peter Straub notes, Wells's refusal to honor distinctions between humans and beasts outraged his critics (xii). Arthur B. Evans writes that early sf critics specifically disliked the genre's lack of emphasis on the human (181). Some sf works, then, are rhetorical in their unique means of casting humanity in a post-Copernican—and by the later nineteenth century, post-Darwinian—light. Like much later sf, some early works hold implications for the place of humans in the universe; that is, they contribute to a reconsideration of an underlying and largely unquestioned anthropocentrism that has dominated human thought across the planet from time immemorial.

Karl Kroeber holds that sf dramatizes "how our world has become so exclusively humanized as to be self-diseased" (22). The antianthropocentric response in sf is not a "niche" at all but rather cuts to the center of the genre. Virtually from its beginnings, wherever that may be (Lucian, Kepler, Voltaire, Shelley, Verne, or Wells), sf shares the idea with conventional nature writing that we are not alone and that the human viewpoint is not the only one that matters. It is commonplace and somewhat justified to think of sf as most regularly locating our neighbors in imagined, faraway worlds, but much sf is as focused on the earth—among our fellow animals, flora, rocks, oceans, and rivers—as is conventional nature writing. Yet only a few of the works I discuss in this chapter may be termed accurately "ecological sf," which is, in some respects, a later category represented by Frank Herbert, Ursula K. Le Guin, and Kim Stanley Robinson, among others.

Since World War II and the dropping of the Atom Bomb utterly changed sf (and virtually everything else), I concentrate in this chapter on sf—mostly American and British—published before that watershed event. My reading of a largely overlooked wing of a popular literary genre will demonstrate that works not often discussed by ecocritics present a view that is fundamental to an ecological conscience: the questioning and even subversion of a hard anthropocentrism. Along the way, I suggest how sf contains ideas worthy of consideration as well as draw attention to works that are nearly forgotten, though many of the texts I discuss are very well known.

Early sf in particular is decidedly patriarchal, since so few women published sf before World War II. But perhaps at least some early sf is more feminine than is commonly acknowledged. It is no coincidence that the decidedly pessimistic views represented in many early sf writers moderated after the Women's Movement and the appearance of such celebrated female sf writers—many of whose works question anthropocentrism—as Kate Wilhelm, James Tiptree, Jr. (Alice B. Sheldon), Sheri S. Tepper, Ursula K. Le Guin, Margaret Atwood, Suzy McKee Charnas, C.J. Cherryh, and Octavia E. Butler, among many others, some of whom I discuss in the next chapter. Darwinian science and its implications for the place of humans in the world didn't vanish, but the prewar naturalistic-modernist strain did give way to a more humanistic understanding of our relationship with the greater world, and this is, at least in part, an effect of the contribution of feminist thought that arose at about the same time as the modern environmental movement.

Science Fiction to the Late Nineteenth Century

In writing about "early" sf, the critic is faced with the question of the genre's beginnings, and a definitive answer is elusive. The genre developed slowly—some of the second century CE Roman writer Lucian's works (see Chap. 2). more or less fit into the genre—and by degrees came into its own with the rise of modern science, culminating, to some degree, with Darwin's theories. Copernicus revolutionized every field of knowledge and all arts; his momentous discovery applies uniquely to the development of speculative literature. So, too, did Copernicus' discovery shake up the anthropocentric view: since the universe does not revolve around the earth, it makes little sense to continue to see man as the center of the universe.

Another founder of modern science, Johannes Kepler, wrote a fictional work, *Somnium* (or, *The Dream*, c. 1611), which is not only a work of fantasy but clearly rhetorical. It describes a trip to the moon as a means of, ultimately, arguing for a heliocentric system. Among his lengthy notes explaining scientific theories and the tale's context, Kepler makes it clear that his work is in opposition to humankind's "ancient Ignorance" (qtd. in Evans 165). Regarding Kepler's *Sominum* and Francis Godwin's 1638 romantic adventure tale *The Man in the Moone*, Arthur B. Evans writes,

> It is not unreasonable to discern in these two "ur-texts" of sf criticism origins of two distinct but interwoven traditions in the history of sf itself: "didactic sf," which is pedagogical by design and gives primacy to scientific exposition over the fictional narrative (e.g., Verne), versus "romance sf," which is more visionary by nature and gives primacy to the fiction over the science—or pseudo-science—embedded within it (e.g., Wells). (167)

Central to the present study is the notion that "romance sf" may also be didactic, scientifically as well as philosophically.

In Chap. 4 I discuss Voltaire's *conte Micromégas: A Philosophical Story* (1752) as a post-Newtonian text, clearly a work of sf, that displaces humankind from the center of the universe, as well as a handful of other non-sf works by Voltaire and other philosophes that operate in a similar fashion. These works, following texts by natural philosophers from the early modern period to the later eighteenth century, formed for the English Romantics the basis for much pre-ecological thought, antianthropocentric speculation, and sf-related themes, including apocalyptic-millennialist rhetoric, found especially in the writings of William Blake and Percy Shelley, but there is little in the English Romantics' work that might be called sf, as opposed to visionary or fantastic. These include works I discuss in Chap. 3: Byron's sf poem "Darkness" and Mary Shelley's *Frankenstein* (1818), which inaugurates a number of themes and even genres, including perhaps modern sf itself. Victor Frankenstein tells his host, Robert Walton, just before relating how he created the monster, "how dangerous is the acquirement of knowledge, and how much happier that man is who believes his native town to be the world, than he who aspires to become greater than his nature will allow" (35). Just before encountering the monster for the first time after its creation, Victor recounts the sublime vistas near Mont Blanc and asks why humans "boast" of being superior to other creatures (75). And much later in the novel, Victor concludes that he cannot comply with the monster's wish that he create a mate for him: "I could not sacrifice the whole human race" (156). Later sf writers, including Shelley herself in her 1826 novel *The Last Man*, would have fewer qualms in dramatizing such a sacrifice.

Leslie Fiedler argues that American fiction, beginning with the gothic novels of Charles Brockden Brown and continuing with *The Scarlet Letter*, *Moby Dick*, *Adventures of Huckleberry Finn* and beyond, are particularly anti-realist in approach (see *Love and Death* Chapter) Although the subject of debate, Fiedler's insight helps explain the nation's early predilection for sf, fantasy, and related forms that shun verisimilitude for romance. Washington Irving's most memorable fiction certainly provides evidence for Fiedler's thesis, as do many of the other American works I discuss in this chapter. Parts of his *A History of New York* (1809) is comparable with Voltaire's *Micromégas* and shares some of its witty tone, though Irving is much more lighthearted and playful. Yet Irving's book is not without serious ideas. His sketch of the picturesque, sublime scenery of the Hudson River area in Book VI, where "the hand of cultivation had not as yet laid low the dark forests, and tamed the features of the landscape" (244), would have a strong effect on Bryant, Cole, and other American writers and artists.

In Book I, Irving satirizes imperialism (with implications for anthropocentrism) as Dietrich Knickerbocker—Irving's historical pseudonym and frame—sets out to "prove" the right of European colonists to take possession of the New World from its native people. Knickerbocker imagines moon men, "possessed of superior knowledge in the art of extermination—riding on hyppogriffs—defended with impenetrable armor—armed with concentrated sunbeams, and provided with vast engines, to hurl enormous moon-stones" (77). They easily take possession of the earth, which is "inhabited by us, poor savages and wild beasts" (78), who differ from the moon men "inasmuch as they carry their heads upon their shoulders, instead of under their arms—have two eyes instead of one—are utterly destitute of tails, and of a variety of unseemly complexions, particularly of horrible whiteness—instead of pea-green." As the Europeans did with the New World's native population, the moon men impose their conception of "the light of reason—and the comforts of the moon" (79) and lead the "infidel savages" of the earth out of the darkness of Christianity to "make them thorough and absolute lunatics" (80). Though Irving is not an early biocentrist, his comic history offers the possibility of a species higher than us.

Among American Romantics (as discussed in Chap. 5), a number of works establish ideas that would receive thorough treatment in later sf. These include James Fenimore Cooper's Leatherstocking tales, which at various points imagine an American wilderness absent Europeans and bemoan its rapacious development. As discussed above, William Cullen Bryant's 1834 poem "The Prairies" suggests the mutability of human civilizations within the context of an enduring earth. Edgar Allan Poe is regularly and correctly regarded as one of the developers of sf. His rich imagination merges with science throughout his oeuvre, from his early "Sonnet—To Science," which (following earlier European models) bemoans the impact of Newtonian science on the romantic imagination, to his final major work, *Eureka: A Prose Poem*, an attempt to close the cleavage between science and literature. Poe's early but important (and oblique) poem "Al Aaraaf" appropriates Muslim accounts of Limbo and the destruction of the earth, and it refers to ruined cities, including Gomorrah (in ancient Canaan; see Gen. 14), Persepolis (in modern Iran), and Tadmor (i.e., Palmyra, in Syria) (see 2.36, 38). Part of the poem's importance lies in its anticipation of his three Platonic/spirit dialogues, "The Conversation of Eiros and Charmion" (1839), "The Colloquy of Monos and Una" (1841), and "The Power of Words" (1845). In the second of the dialogues, humanity brings ruin upon itself, at least indirectly. Monos tells his feminine double Una that, through a lust of the intellect and the destruction of the natural world, we "worked out our own destruction." Monos states that the leading "evil" is knowledge: "Man could not both know and succumb. Meantime huge smoking sites arose, innumerable. Green leaves shrank before the hot breath of furnaces. The fair face of Nature was deformed as with the ravages of some loathsome disease" (610). Poe raises a paradox, introduced in "Al Aaraaf," that would be confronted by such twentieth-century sf works as Walter Miller's *A Canticle for Leibowitz* and Philip K. Dick's story "Autofac" (discussed below) in regarding knowledge both as inevitable in civilizations and potentially the ultimate cause of the end of our species.

Although Poe disliked the New England Transcendentalists, who were, he believed, not only overly optimistic but overly didactic, through Monos, he posits not un-Emersonian notions about the uses of "analogy which speaks in proof-tones to the imagination alone." Monos echoes Rousseau and anticipates Thoreau, the forefathers having pondered "the ancient of days when our wants were not more simple than our enjoyments were keen" (609). The story, writes Ottavio M. Casale, carries "a Thoreauvian distrust of materialism to apocalyptic extremes, attributing a projected destruction of the world to lust of intellect and the ravaging of nature" (365). Yet with regard to nature, Poe has much more in common with Thomas Cole, another skeptic of democracy whose work is infused with antidevelopment rhetoric. Out of a foolish belief in the supremacy of "the Arts," Monos says, "Man . . . could not but acknowledge the majesty of Nature, fell into childish exultation at his acquired and still-increasing dominion over her elements" ("Colloquy" 610). Before Monos proceeds into the story's second half, which concerns his own premature burial, he decries the world's overreliance on "harsh mathematical reason": "This the mass of mankind saw not, or, living lustily although unhappily, affected not to see. But, for myself, the earth's records had taught me to look for widest ruin as the price of highest civilization" (611). As a catastrophe story, "The Colloquy of Monos and Una" anticipates a wide range of later works that dramatize human-caused apocalypse resulting from unwise use of the environment or the out and out ravaging of the earth for private gain.

Poe's first spirit dialogue, "The Conversation of Eiros and Charmion," revolves around Eiros's description of the destruction of the earth by comet to Charmion (both names are variations on characters' names in Shakespeare's *Antony and Cleopatra*). Rooted in the biblical idea that the world will be destroyed by fire, the story is also filled with pseudoscientific details on comets that may be a reply to the popular 1839 story "The Comet" by S. Austin, Jr. (see Mabbott's commentary 452–53). Poe was, like many writers of his time, fascinated not only with biblical apocalypse but with fallen empires. A picture of the mindset of American romantics of the era, his early poems "The City in the Sea" (1831) and "The Coliseum" (1833) were written around the time Cole, like Poe a Byron and Volney enthusiast, speculated on America as future ruins while in Sicily (see L. Ferber 168).[1]

Of his major works, *A Connecticut Yankee in King Arthur's Court* is Mark Twain's major contribution to sf. The novel praises machines and Yankee ingenuity, though in the non-sf *Adventures of Huckleberry Finn* and elsewhere, Twain expresses disdain for a world advancing toward mechanization, capital, and civilization (even as Twain would invest heavily —and lose a fortune—in a typesetting machine). In Chap. 19, as Huck and Jim float in pastoral calm down the Mississippi River, removed from human civilization, they are at ease. As Lionel Trilling notes, Huck's character is defined by the few quiet nights on the river (314). By daylight, however, Huck's serenity amidst the "solid lonesomeness" of the river is interrupted by signs of what Twain would later call "the damned human race" as he observes "a wood-yard, likely, and piled by them cheats so you can throw a dog through it anywheres" (96)—a subtle but clear representation of human wickedness in contrast to the serenity of nature. A page or so later, Huck and Tom will meet the novel's principal confidence men, the duke and dauphin, and encounter cheats first hand. As a satirist, Twain never states directly that anthropocentrism is a false position. Allowing the perspective of a naïve narrator to reveal human greed, Huck almost never judges or condemns judges the crooks and swindlers (including his own brutish father) throughout the novel.

In his later works, culminating in *The Mysterious Stranger*, which is closer to fantasy than sf, Twain's pessimism toward humanity would be even more pronounced. Some of Twain's posthumously published non-fiction, collected by Bernard DeVoto in *Letters from the Earth*, betrays as negative a view of humanity as exists in the work of almost any major writer. The title essay veers closely to the Voltaire of "Plato's Dream" in attacking man's hubris in claiming he was made in God's image. In fact, God says, "Man is an experiment, the other animals are another experiment. Time will show whether they were worth the trouble" (6–7). In a piece entitled "Was the World Made for Man?," Twain, voiced thinly as a "scientist and theologian," lets loose with full force on anthropocentrism. The piece concludes:

> Man has been here 32,000 years. That it took a hundred million years to prepare the world for him is proof that that is what it was done for. I suppose it is. I dunno. If the Eiffel Tower were now representing the world's age, the skin of paint on the pinnacle-knob at its summit would represent man's share of that age; and anybody would perceive that that skin was what the tower was built for. I reckon they would, I dunno. (226)

Later in life Twain grew more pessimistic and thought more deeply about Darwin, but he maintained distance, and this is the key to his art, early and late. In the first volume of his *Autobiography*, discussing compliments, Twain states that he appreciates a small number of them, but the thought of adulation sends him into a misanthropic tirade:

> What a king must suffer! For he knows, deep down in his heart, that he is a poor, cheap, wormy thing like the rest of us, a sarcasm, the Creator's prime miscarriage in inventions, the moral inferior of all the animals, the inferior of each one of them in one superb physical specialty or another, the superior of them all in one gift only, and that one not up to *his* estimation of it—intellect. (184)

With Twain among the great American pessimists of the late nineteenth century, Ambrose Bierce wrote a handful of works that might be called sf, though most of these are not among his most memorable endeavors. His story "Moxon's Master," in which is a chess-playing automaton apparently murders its creator, is cited regularly as one of the first robot stories. "For the Akhoond" is a sf story that combines apocalypse and a vaguely defined ecological theme. Reflecting national politics in the late nineteenth century, in the forty-sixth century, the former United States is divided into Galoots, Pukes, and Smugwumps. The latter of these had pretentiously held the "vain delusion that they could subdue Nature" (177), but they were wiped out by cataclysmic climate change. The extinct inhabitants of the southern part of the nation—designated by the problematic designations "Crackers and Coons"—were succeeded by savage beasts, reptiles, and "offensive" birds, though, as the narrator states, they had "always been more or less dead" (179).[2] Bierce's poem "A Vision of Doom" (1892) is comparable with Poe's spirit dialogues and Byron's "Darkness"; as in Poe, the poem is delivered by a posthumous consciousness, whereas in Byron, the narrator apparently speaks from his tomb. Yet the poem is a sort of augury for doom, a time foreseen by a "Great poet" when "that foul city be no more!" In the end, reflecting an idea that goes back at least to Ovid, greed does humankind in: "The people gathered gold, nor cared to loose/The assassin's fingers from the victim's throat" (p. 48).

H.G. Wells and After

The implications of Darwin's *On the Origin of Species* would have a far-reaching effect on the development of sf in the later nineteenth and early twentieth century, just as Copernicus and Galileo's astronomy influenced Neoclassical/Enlightenment writers in the seventeenth and eighteenth centuries. The worldview shifted even further away from an anthropocentric one, in thought if not in action. Yet it is a mistake to read every *fin de siècle* sf (or speculative) work as a pure Darwinian parable. Richard Jefferies's best-known work, *After London* (1885), owes less to Darwin than to eighteenth-century British writer (and major Thoreau influence) Gilbert White, though the opening chapters of the novel reflect a dark shift in speculative literature representative of the era. The novel's opening paragraph frontloads the novel's context: "The old men say their fathers told them that soon after the fields were left to themselves a change began to be visible. It became green everywhere in the first spring, after London ended, so that all the country looked alike" (11). Jefferies provides a pseudo-scientific cataloguing of England in the year 3000 after a vaguely-described catastrophe has all but extinguished humanity, the remaining Britons of which have resorted to feudalism. The sea has risen, flooding London, and the Thames is nothing more than a poisonous swamp. Jefferies describes the altered ecological chains: crops are eaten by mice, which are eaten by hawks; cats and dogs go feral (15, 18).

Post-catastrophe, the narrator celebrates this natural, largely human-less world through minute details. Unfortunately, the human romance that proceeds after the early chapters doesn't sustain the novel. As Brian Aldiss writes, Jefferies's "people enjoy less animation than his plants" (114).

The early writings of H.G. Wells are central in this chapter. They are among the most widely-read works I discuss, and they have left an indelible mark on Western culture. Most of Wells's best fiction centers on shaking up well-entrenched views of humans about the world, including a thorough questioning and subverting of anthropocentricism within an ecological framework. Whereas later Wells works tend toward the utopian, his earlier scientific romances are dystopian, pessimistic, though Wells was early on aware that evolution is "no mechanical tendency making for perfection . . . it is simply the continual adaptation of plastic life, for good or evil, to the circumstances that surround it" ("The Man" 5). When the Time Traveller of *The Time Machine* (1895) arrives in the 803rd century, he has "happened upon humanity upon the wane," and any vestiges of anthropocentrism are a mockery. Humanity had, over the millennia, civilized the world—"One triumph of a united humanity over Nature had followed another" (29)—but an era of stagnation ensued following the "defeat" of nature. There are no more gnats, weeds, or fungi, but "after the battle comes Quiet" (31), and humans—or at least the (once) capitalistic surface dwelling Eloi—have become soft, lazy, and ignorant. The triumph, remarks the Traveller, is "too perfect" (32).

Evoking the romantic symbol system from a few generations before in writings by Madame de Staël, Chateaubriand, Byron, and others, Wells includes several ruins images in *The Time Machine* to underscore the mutability of human civilizations. Writing less about the future than about the present and immediate future, the novel serves as a warning about inequalities (the growing chasm between the rich and poor) and the unchecked human exploitation of the earth. Wells presents the idea more explicitly in his story "A Dream of Armageddon" (1901), the internal narrator of which comments on the builders of warplanes, which will lead eventually to the Armageddon of the title: "You know the silly way of the ingenious sort of men who make these things; they turn 'em out as beavers build dams, and with no more sense of the rivers they're going to divert and the lands they going to flood!" (187). Between the lines of such statements is Wells's famous maxim at the end of *The Outline of History*, "Human history becomes more and more a race between education and catastrophe" (1100). The Time Traveller observes the logical conclusion, after hundreds of centuries, of the political system of England, the aftermath of the triumph of capitalists over both nature and fellow human beings: visiting the far future, the Traveller finds a silent, darkened earth—without humans. After a long interval, as he travels backwards to the present, marking eons of time, he at last sees "the evidences of decadent humanity" (81).

From almost the beginning of *The Island of Dr. Moreau*, Wells's protagonist, Edward Prendick, is placed in a world in which the demarcation between humans and animals is unstable. The novel may be read as a wry take on the English class structure, an antivivisectionist parable (echoing Victor Frankenstein's warning against scientific hubris), and an adventure tale, but the blurring of human-animal distinctions is central. By Chap. 12, horrified by scientist Moreau's experiments in creating half-man, half-animal creatures ("Beast Men" or "Beast People"), Prendick realizes that he and the Beast Men share more similarities than he cares to admit. He ridicules their primitiveness, but drilled in their laws ("Not to go on all Fours," "Not to suck up Drink," etc.), Prendick's "I" and the Beast Men's "they" begin to merge in a ceremonial chant: "A kind of rhythmic fervor fell on all of us; we gabbled and swayed faster and faster, repeating this amazing law. Superficially the contagion of these brute men was upon me, but deep down within me laughter and disgust struggled together" (80). Prendick follows these creatures' example; that is, he *apes* them.

The monstrously hubristic Moreau is the novel's true "beast." He should know better than to tamper with nature's laws. But it is his sheer anthropocentric arrogance that is the root of his terrible experiments. It has caused him to learn the wrong lessons: says Moreau, "The study of Nature makes a man at least as remorseless as Nature" (102). He recounts how, having first laid eyes on the island where he conducts his vivisections, "The place seemed waiting for me" (103). Prendick, trained (like Wells) by T.H. Huxley at the Royal College in Kensington, discovers firsthand that species are mutable. After several weeks among the Beast People, he "fell in with their persuasion that my own long thighs were ungainly" (113). Unlike Gulliver after returning to England from the land of the rational, horse-like Houyhnhnms, he does not regain confidence in humanity or come to terms with his own humanness; having returned home, the streets of London, Prendick notes in the final few paragraphs, are filled with "prowling women" and "craving men" (183). He undergoes no reaffirmation.

At first glance, *The War of the Worlds* (1898) would seem to have little in common with its predecessors, but its biological, detached view of the invaders from Mars correlates with the Time Traveller's discoveries about the finiteness of humanity and the accentuated blurring of the line between human and nonhuman in *Moreau*. *The War of the Worlds* reveals, again, Wells's instruction in biology under Huxley. Reflecting the indifference of nature, Wells applies natural selection and speculative science to the invading Martians themselves. Beginning with the book's epigraph (from Kepler: "how are all things made for man?" (2)), Wells uses the Martian attack as a means of questioning a hard anthropocentric viewpoint. The novel's opening paragraph has the Martians scheming to invade earth, looking at humans the way that we look at "transient creatures" through a microscope (7), and two paragraphs later he attacks directly the vanity that blinds humans (8). The oft-filmed novel retains its powerful ability to shake up our assumption that the earth exists for human benefit, mainly. Humans are fully capable of being exterminated by outer forces (Martians, for example), just as sailors exterminated "the respectable dodo" (34).

Further, *The War of the Worlds* forces a reexamination of the relationship of humans toward their fellow earthly inhabitants, as well as England's late Victorian colonizing tendencies (also a theme in *Moreau*). As the Martians rapidly gain control of England, institutions that the people have taken for granted such as the police and the railways are "quickly losing shape and efficiency, guttering, softening, running at last in the swift liquefaction of the social body" (92). The veneer of civilization, Wells demonstrates, is thin: the narrator and other human survivors are reduced to mere scavengers; "how swiftly that desolating change had come" (147). The narrator has become "like a rat leaving its hiding place" and "an inferior animal." If the invasion has taught humans nothing else, it "has taught us pity—pity for those witless souls that suffer our dominion" (149). The narrator's description of the vampiric means of Martian feeding is admittedly "repulsive," but he includes a caution against anthropocentrism: it is only repulsive from a human point of view. Sovereign man has become mere Martian food. Just as humans have transformed the landscape and run rabbits out of their homes, so have the Martians transformed the landscape with their curious red flora. Underscoring human helplessness but also affirming humans as a part of the natural order, the invaders are slain not by modern British weaponry but by "the humblest things that God, in his wisdom, has put upon this earth"—bacteria (168).

Many of Wells's short stories also reflect the author's Darwinian rejection of anthropocentrism. His 1897 apocalyptic story "The Star" envisions the events before and during the collision (or, as it turns out, near-collision) of an unnamed star (planet) with the earth. "The Star" echoes Poe's "The Conversation of Eiros and Charmion," though Wells's story operates not through dialogue but a series of vignettes of people in various stations responding to the rapidly approaching star. Against the current of opinion, a "master mathematician" makes calculations and concludes, "Man has lived in vain" (284). As the star reaches full impact, the final paragraph transposes to "Martian astronomers—for there are astronomers on Mars," who note little change in the appearance of the earth's surface, unaware that the passing star has all but exterminated human life. The story's concluding idea is characteristically Wellsian: the Martian astronomers' indifference "only shows how small the vastest of human catastrophes may seem, at a distance of a few million miles" (289). (Wells's concluding passage echoes Mary Somerville, quoted above: "the whole system of which [the earth] forms so small a part,—might be annihilated, and its extinction be unperceived in the immensity of creation" (4).) "The Empire of the Ants" (1905) initiated a sub-genre of sf centered on malignant, imperialistic insects—a few inheritors of which I discuss below. The meticulous British engineer-narrator is uneasy with the lack of order in Brazil, a country he finds characterized by an "inhuman immensity" (588). In England, things "grow on lease," but here, humans are not in charge fully. Over time, humans evolved and took charge of the earth; what is there to prevent the invading hordes of ants, which have an intelligence and language of their own, from doing the same? (589).

Wells would, beginning in the 1900s, reposition himself as a utopian visionary, but his final word, stated in *Mind at the End of Its Tether* (1945), is that all of life—and not just human—is "played out" (18). A reflection, doubtless, of his own ill health (he would die the following year) and the wearying strain of the second world war, Wells gives no specifics in explaining his pessimism, other than to echo Darwin on "the struggle for existence." He holds a shred of hope that *Homo sapiens* ("as he has been pleased to call himself" (18)) may adapt, but, he concludes, "my own temperament makes it unavoidable for me to doubt, as I have said, that there will not be that small minority which will succeed in seeing life out to its inevitable end" (34).

Like Wells, Matthew P. Shiel was interested in science, but his narrative style, in *The Purple Cloud* (1901), replete with plenty of "purple" passages, looks back to Poe. This novel suggests ecological themes that bear comparison with nineteenth-century "last man" stories. Returning by boat from a long, fierce but successful attempt to arrive at the North Pole, Adam Jeffson finds a dead world on the ocean and reflects on his mood: "that abysmal desolation is loneliness, and sense of a hostile and malign universe bent upon eating me up: for the ocean seemed but a great ghost" (101–02). A bit later, suggesting a vaguely biocentric viewpoint, he observes porpoises leaping in the ocean and says, "I am not quite alone": the sea "had its tribes to be my mates" (105). On land in Dover, he says, "Well, Lord God, Thou hast destroyed the work of thy hand" (130). The mysterious purple cloud—a "slow-riding vapour which is touring our globe"—has destroyed almost all living things, despite the "sparing nature" he has witnessed previously. His religious faith wavers, but he quotes Job, "Though He slay me, yet will I trust in Him" (139). In Kent, he notes "an aspect of return to a state of wild nature"—soil untouched for at least a year (152). Then he notes the "exuberance" of the natural scene owing "to some principle by which Nature acts with freer energy and larger scope in the absence of man" (153). An effect of the passing of humans, storms, earthquakes, and the like grow worse—an idea akin to some old philosophies, including those of Lucretius and Seneca. Jeffson wonders if the earth is "herself a living being, with a will and a fate" (277); the earth, he concludes in his brooding reasoning, is conscious of the extinction of man: great is the earth, and the Ages, but man "passeth away" (279).

Jeffson has some knowledge of the earth, but its means of "forming coal, geysers and hot sulfur-springs, and the jewels" and other phenomena are mysteries—"I do not know them, but they are of her, and they are like me, molten in the same furnace of her fiery heart" (278). Out of his enormous will to power, he proceeds, in an ambiguous, Orlando-like gesture of his own survival, to burn the great cities of the earth he visits, all the while arrayed in a Byronic oriental dressing. "[H]ere I am Sole; Earth acknowledges my ancient sway and hereditary scepter: for though she draws me, not yet, am I hers, but she is mine" (302–03). In a forest somewhere in the Middle East, he observes, "Here nature in only twenty years has returned to an exuberant savagery" (326). To his mate Leda—his "Eve"—Adam speculates that humanity will at least attain "the nobility of self-extinction" (367). Near the Balkan Mountains, he states that it is "wonderful to see the villages and towns going back to the earth, and already invaded by vegetation, and hardly any longer breaking the continuity of pure Nature, the town now as much the country as the country, and that which is not-Man becoming all in all with a certain *furore* of vigour" (405). Adam initially chooses annihilation and refuses his Eve, but he reneges, falls in love with her, and sets his sights toward propagating the species.

Seven years later appeared William Hope Hodgson's *The House on the Borderland* (1908), which is not ecological and contains no social commentary but is effective in showing a dying and dead world long after humans have disappeared. Exploring images Wells set down in *The Time Machine*, Hodgson is more drawn toward the horrible and weird than science. An earlier inheritor of the title house witnesses the Hindu goddess of death, Kali, and the Egyptian Set, the soul destroyer, as he is transfixed out of human time to experience "the immutable, awful quiet of a dying world" (99). As the sun begins to transform to a "vast dead disk" (101), a flame cuts across the narrator's darkened vista; a "glare that lit up the dead earth, shortly; giving me a glimpse of its flat lonesomeness" (102), and "life and light, and time, were things belonging to a period lost in the long gone ages" (103). Hodgson would visit the theme again in his longer, haunting novel *The Night Land*. It is easy to see how Hodgson would bear a strong influence on later writers of the weird, including H.P. Lovecraft, who was also interested in posthuman narratives.

Early Twentieth Century

Although Wells was hardly the first writer to speculate on the far future destiny of the earth, he (with Jules Verne) established this frame as a norm in sf. The pessimistic idea of a degraded far future humanity does not in itself constitute a questioning of or attack on anthropocentrism, but this is often the case. In his 1960 survey of sf, *New Maps of Hell*, Kingsley Amis writes that the genre "presents with verisimilitude the human effects of spectacular changes in our environment, changes either deliberately willed or involuntarily suffered" (26). Amis notes sf's recurring themes of our need for but lack of security and speculation about our own extinction (69), but he finds "the activist temper of SF reassuring" (131), appreciates the genre as a "medium in which our society can criticize itself, and sharply" (155) and "make some contribution to the security of our" future (157).

All these ideas apply to E.M. Forster's influential story "The Machine Stops" (1909), which is a pessimistic response to Wells's later utopias. The story does not concern the extinction of the human race per se, but its world is utterly anti-utopian. Anticipating the century's three most influential dystopian novels—*We*, *Brave New World*, and *1984*—humans are shut out completely from the natural world and are raised, sustained, entertained, disciplined, and governed by the machine. There is no indication that a great cataclysm has occurred; life as it is has arisen over time through a gross overdependence on machinery. At the end of the story, as the title of the story is realized through internal malfunction, the main character, Vashti (a middle-aged woman), and her fellow machine dwellers, the majority, face a sort of apocalypse, but it is one that might ultimately be a good for humans: "Homeless" people have been living outside the machine, in the natural world, all along. Vashti represents the status quo, those unquestioningly content living in the machine. She scarcely knows her own son, Kuno, limited to brief correspondences with him through the machine—difficult for contemporary readers to not identify with the Internet. She mistakenly tells Kuno that the earth's surface is uninhabitable, lifeless, and composed only of dust and mud (149).

The world of this story is one of thorough uniformity: the machine life in Asia is exactly the same as the machine life in Western Europe, though the centralized brain of the machine insures that nationalities are kept alive. Air-ships deposit people around the globe, and few travelers desire to even look outside, much less venture outdoors. Even the stars are "intolerable" (158–59). Thus, "All the old literature, with its praise of Nature, and its fear of Nature, rang false" (156). With this utter hatred of the outdoors is contempt for its use beyond commodity. So the forests to the north of the Himalayas had been "destroyed during the literary epoch for the purpose of making newspaper pulp" (162–63). Only homeless dissidents such as Kuno realize that through annihilating space, "We have lost a part of ourselves" (167). Once out of the machine, Kuno tells Vashti he saw "low colorless hills. But to me they are living and the turf that covered them was a skin, under which their muscles rippled" that had once been possessed of "incalculable force" to enlighten people, who are dying because of their isolation and their dependence on the machine. As the machine stops at the end of the story, Vashti realizes "that civilization's long day was dying" (194).

Jules Verne is, of course, one of the founders of sf, but it is his very late story "The Eternal Adam" (1910), written partially by the writer's son Michel, that concerns us here. The fact that Verne grew slowly more pessimistic over time is reflected clearly in this story, which though not particularly "ecological" is both apocalyptic and anti-teleological. As the story, set in the far future, opens, the main character, Zartog Sofr-Ai-Sr, is walking toward a celebration of the 195th anniversary of "The Empire of the Four Seas," which extends from present-day Berlin to Cape Horn. Sofr, a philosopher-scientist, recalls the long, slow movement humankind has accomplished over those centuries as it at last "freed itself from its bestial origin" and succeeded in the "taming of material nature" (213). He accepts evolutionary theory but refuses to consider the possibility, theorized about of late, of a sophisticated society some 40,000 years before. Sofr is reluctant to admit that ancient humans would leave no trace of a civilization comparable with or superior to that of the present. This, he states, "would be to deny the future, to announce that our efforts are all in vain, and that all progress is as precarious and as uncertain as a bubble of foam on the surface of the waves!" (219). In the midst of these thoughts, Sofr personally uncovers in a hidden site a journal, the contents of which reveal a historical *mise en abyme* that unravels Sofr's confidence in human achievement.

The journal, 20,000 years old, was written and narrated by an early twentieth-century Frenchman. The beginning entry establishes a sense of urgency and his purpose in describing "those frightful happenings . . . for the enlightenment of those who come after me—if indeed mankind is still entitled to count on any future whatever" (222). In a "revolt of matter," a devastating earthquake strikes which causes the earth's land to sink and its water to rise. Escaping by automobile, the main character, with a few others, ascends to higher land, while all the land surrounding him submerges and disappears. A ship, *The Virginia*, rescues the main character, and after months of sailing the world on a seemingly endless sea, he succumbs to the realization of "loneliness in the midst of a pitiless universe" (233). At last, the ship finds a stark, inhospitable piece of land, where the survivors live on turtle eggs and seaweed, until, years later, they are able to grow wheat. Humankind falls rapidly into a state of retrogression. All human learning disappears, and sustenance becomes the sole aim. The inhabitants live nakedly, clothes being too difficult to make and maintain. The journalist grows old and the ancient document concludes; Sofr becomes convinced of the Pythagorean "eternal recurrence of events" (244).

Another tale of the future, though not distant, Jack London's *The Scarlet Plague* (1912) functions partly as a piece of socialist-naturalist rhetoric and partly as an ecological novel. After the Great Plague of 2013, which wiped out most of the world's human population, an old man—a former English professor—tells his grandsons the story of how the plague occurred and the events that followed. In a fiercely anti-intellectual post-plague world, only the man, Ganser, retains some of the old knowledge. To his impatient audience, Ganser mumbles a part of a poem —"fleeting systems lapse like foam . . ." (unidentified but from *The Testimony of the Suns* (1902), by London's friend George Sterling)—and concludes, "That's it—foam and fleeting. All man's toil upon the planet was just so much foam. He domesticated the serviceable animals, destroyed the hostile ones, and cleared the land of its wild vegetation. And then he passed, and the flood of primordial life rolled back again, sweeping his handiwork away" (11).

As in Jefferies's *After London*, the absence of humans allows wildlife to flourish. Wolves reappear on Cliff House Beach in San Francisco (the setting of Ganser's storytelling), the grizzly population increases, and the now-thriving mountain lions have driven horses to the beaches, while the "the sea lions, bellowing their old primeval chant, hauled up out of the sea on the black rocks and fought and loved" (63). Such passages involving the fate of domesticated animals and cultivated land post-plague anticipate the more successful George R. Stewart 1949 postapocalyptic novel *Earth Abides* (see next chapter). London's plague is a social evener, as much as it is an ecological resetting, post-human. Yet Ganser suggests cyclical history in predicting the human race's fall into primitivism (signs of which are apparent in all but one of the grandsons) and its inevitable "bloody climb upward to civilization" (14).

Though flawed, London's novel is enlightened in its sense of ecological order. Much the same could be said of Murray Leinster's 1920 novelette *The Mad Planet*, the scenario of which is startlingly contemporary: Due to high amounts of carbon dioxide in the atmosphere, global warming develops, killing off humans except for those living in higher elevations. By the middle of the twenty-first century, the remaining humans fall into near savage conditions and fight over the last habitable places on earth. Most of the earth's trees die out, and humans, now fewer in number, begin to adapt to new conditions. Leinster's world of the far future is ruled by giant, threatening insects that buzz through the air that resemble aircraft flying over modern cities. The pink, puny humanlike hero of the story, Burl, proceeds to turn humanity toward hunting again, slays giant insects in combat, woos his sweetheart Saya, and puts humankind back on the path to civilization, all in a few hours.

Published a year later, J.D. Beresford's "A Negligible Experiment" dramatizes the approaching destruction of earth by an approaching planet larger than Jupiter. The narrator's fellow observer speculates (echoing the concluding paragraph of Wells's "The Star"—Wells was a heavy inspiration for Beresford) that beings in a far reach of the universe "may catch sight of this tiny blaze of ours—and wonder. It will be relatively a very small affair." As in Poe's spirit dialogues, our spirits may continue to exist, but, the narrator concludes, "we are the creatures of some chance evolutionary process, or we are an experiment that has failed" (152–53). In Beresford's 1929 story "The Man Who Hated Flies," the title character uses an insecticide that wipes out flies successfully but also obliterates many pollinators of plants (see Stableford). E.B. White's celebrated 1950 story "The Morning of the Day They Did It" follows a similar theme.

In another part of the world, Yevgeny Zamyatin both operated in and criticized the policies of the Communist Party of the Soviet Union after the 1917 Revolution, though he had been (like Lenin, Stalin, and Trotsky) a committed member of the Old Party Guard. His novel *We* (1921) was quickly banned by the Soviet censorship board and was smuggled to the West. It is not explicitly antianthropocentric but holds implications for the mechanization and industrialization of the natural world and would be a heavy influence on later writers. *We* is set in an unspecified future date (the twentieth century is, in the novel, antiquity). The natural world is "irrational" because it is nonmechanical. Predating Orwell's *1984* (which was inspired by Zamyatin) by more than two decades, all human activities, including sex, are controlled by the One State (the "we" of the title is "my" in Russian), and any trace of romanticism (widely defined, including even use of the imagination) is forbidden. Any subversives are subject to the "Operation" (the cauterization of the brain) by the dictatorial Benefactor and his Guardians. Looking back to Dostoevsky ("The Grand Inquisitor" in particular), "freedom" is in fact a curtailment and is equated with mere compliance and praise for the Benefactor, the wonders of technology, and scientific rationalism. In this world, "everything is finite, everything is simple, everything is calculable" (201). The subjects are accustomed to the "flawless" skies, and there is no taste for picturesque irregularity (5). Nature has been channeled, and there are no more natural catastrophes (22); there are not even thunderstorms (125). Nature is kept out behind the State-built Green Wall, which, according to State doctrine, is "the greatest of all inventions," to isolate the "perfect, machined world . . . from the irrational, chaotic world of the trees, birds, animals" (83). Everything is "steeled," even the trees, "twisting in convulsions" (42). The self-conscious but naïve and compliant narrator, D-503, has been, as he explains in a footnote, conditioned to see nothing beautiful in flowers, "and the same thing goes for everything that belongs to the wild world, which was chased off long ago beyond the Green Wall. Only the rational and the useful are beautiful machines, boots, formulas, food, etc." (44). Through the cloudy glass of the Wall, D-503 comes face to face with an animal with yellow eyes (probably a cat of some kind, though in this world animal names are forgotten) and wonders, "what if he, this yellow-eyed being—in his ridiculous, dirty bundle of trees, in his uncalculated life—is happier than us?" He concludes that the idea is ridiculous; the animal may be happier than himself, but this is an exception since, he says, he is sick (83).

Unknown to the city dwellers, life outside the Green Wall continues, and it is a realm that represents and constitutes, as suggested later in Orwell and Huxley, rebellion and an unimpeded spiritual freedom. Later in the novel, finding himself behind the wall with his lover, the rebellious I-330, and a group of naked, hairy subversives who succeed in blowing up the wall, D-503 finds it difficult to maintain walking balance on the unlevel, non-concrete earth, and he is deafened by the silence and chokes on the fresh air (136). Before long, though, D-503 finds himself beginning to think not as a component of a great machine but as an individual (138), and he observes that a bird passing by is "alive, like me. It turns its head to the right, to the left, like a person, and then screws its black round eyes into me." The natural environment causes I-330 and D-503 to smile and feel that "the Earth, drunk, joyful and light, is floating . . ." (139). The novel ends with apparent pessimism, though not without the realization, in the wake of an attack on the Wall, that "the greatest and most intelligent civilization in all history was collapsing, but down here, by some irony, everything had stayed like it was: splendid. And to think: all this is condemned, all this will grow over with grasses, and there will only be 'myths' about all this . . ." (200).

As with *We*, its greatest Western counterparts, Aldous Huxley's *Brave New World* (1932) and George Orwell's *1984* (1948), do not critique anthropocentrism directly, but they address fears about the growing isolation of humans from nature. Like *We*, Huxley's world, a futuristic London, is one in which society controlled by a hubristic technocracy consciously cultivates a hatred of nature. As the Director of Hatcheries (which oversees artificial human reproduction) explains near the beginning of the novel, "A love of nature keeps no factories busy. It was decided to abolish the love of nature" (23). A minor but representative character says that mountains are "revolting" (96). Having traveled by rocket to America, psychologist Bernard Marx, who hates the conformity of his world, with his girlfriend, the promiscuous, representatively shallow vaccination worker

Lenina Crowne, find that the Santa Fe area is powered by the "Grand Canyon hydroelectric station" (101)—a bad joke or nightmare to any preservationist-minded reader. She is uncomfortable on a tour of the primitive pueblo Malpais to which Bernard has brought her; she says, "I hate walking. And you feel so small when you're on the ground at the bottom of a hill" (108). There they meet the savage John, more or less the novel's hero, who finds freedom in the "solitude," a concept civilized people do not understand and hate. Later in the novel, John admires the British countryside: "Flowers and a landscape were the only attractions here. And so, as there was no good reason for coming, nobody came" (246). Living in a wild area in an old lighthouse, John seeks independence by farming and hunting small game, and the work gives him "intense pleasure" (247), but when he is robbed of his solitude, he hangs himself.

The world of Orwell's *1984* is similarly homogenized and cleansed of wildness, which carries too strongly for the authoritarian Party the stench of freedom. Protagonist Winston Smith notes that the world is not so much cruel as it is barren, dingy, and listless, since the Party has set up "a world of steel and concrete, of monstrous machines and terrifying weapons" (65). In this world, "The birds sang, the proles sang, the Party did not sing" (196). (The proles—the proletariat—are the great majority of citizens with no political influence living in Oceania, the fictional world of the novel comprising the UK, the Americas, Australia, and other regions.) Human knowledge has willingly, consciously gone backwards. "We make the laws of nature," says inner-Party member O'Brien (236). "The earth," he says, "is the center of the universe" (237) and "[t]he so-called laws of nature were nonsense" (248).[3]

Far removed from twentieth-century political dystopia fiction, H.P. Lovecraft is well known for his "weird" fiction, but some of his stories also meet the criteria for sf. One of these is "The Colour out of Space" (1927), which has long been regarded as one of the writer's best, including by Lovecraft himself. While it is a stretch to call the story ecological, it does have ecological implications, as its narrator is a surveyor for a new reservoir to be built near Arkham, Massachusetts, the fictional town that is the setting for many of Lovescraft's stories. S.T. Joshi quotes a 1935 letter Lovecraft wrote about the melancholy mood that struck him regarding the "vast amount of rural territory [that] was flooded in 1926 for a reservoir in Rhode Island" (Lovecraft 60 Note 6). After a meteor crashes into the Arkham area, life around it changes. In what could serve as a sort of allegory for blighted nature (and, perhaps, genetically modified food), a "stealthy bitterness and sickishness" infects fruit from the area that formerly tasted sweet. Initially, small mammals begin to act strange and mutate while the flora assumes strange colors and shapes, but, suggesting the fact that people are connected inextricably to their natural environments, after a while people begin become ill, including the tragic Gardner family, whose spring has been poisoned by the alien substance. The narrator awaits the building of the reservoir, but he wonders if the blight will eventually seep into it, poisoning Arkham and its surroundings (99). Familiar with Darwin's theories, likely through reading science popularizer Hugh Elliot and Darwin disciple Ernst Haeckel, Lovecraft developed a strongly materialistic viewpoint in which humans are almost nothing in the cosmic scale. Regarding Elliot's denial of teleology, writes Joshi, Lovecraft was passionate: "His cosmicism, engendered by his astronomical studies, had relegated the entire history of the human race to an inessential nanosecond in the realm of infinite space and time" (205). With these thoughts in mind, Lovecraft consciously designed the alien matter carried by the meteor that destroys the land around Arkham as completely nonanthropomorphic and an entity wholly outside the realm of the very limited abilities of humans to decipher.

Lovecraft's work was published heavily in early pulp sf magazines, which provided an audience (and meager paychecks) for a large number of writers, including Edmond Hamilton. In his story "The Earth-Owners," which appeared in a 1931 edition of *Weird Tales*, strange, giant clouds appear over modern Boston. One character, Randon, following the theories he read in a book, posits that the clouds are composed of the earth's "owners," who have returned: "We lords of creation, we humans who dominate the other animals of earth so completely, how few of us have ever dreamed that perhaps our earth and ourselves are owned as completely as we own a game-preserve and its animals!" (22). As Randon's argument begins to sway his companions, and as the cloud descends, against the wind, over the city, killing everyone in its shadow, another character, Carter, says, "What a jolt it would be to humanity to find out that its planet was owned as we might own a farm!" (24). As in *The War of the Worlds*, the clouds, in fact "organic gases" (30), suck the life out of humans: the "earth-owners" have returned merely to feed on humans. Yet without the clouds watching over the earth, Randon speculates, it would have surely been invaded long before by other aliens (138). In the end, it is an alien race that "owns" the earth and battles and destroys the clouds, which have been humans' unsuspected protectors. Although absent of ecological ideas, Hamilton's story not so subtly challenges anthropocentrism.

One of the most important sf pulp magazine editors, John W. Campbell, studied physics at M.I.T. and began writing stories about the far future, some of them under the name Don A. Stuart. His "Twilight" (1932) may be read as a warning against excessive mechanization. An insurance salesman picks up a time-traveling hitchhiker from 1000 years in the future who has traveled another million years into the future. He has observed a solar system populated with large-headed humans who are so "perfected" that (echoing Wells's Eloi as well as Forster) they can no longer do anything for themselves—the machines they designed do everything for them. Humans have forgotten their own technologies and no longer hold curiosity for anything but the machines. The loneliness of future humanity, says the time traveler,

> was beyond hope. For, you see, as man strode toward maturity, he destroyed all forms of life that menaced him. Disease. Insects. Then the last of the insects, and finally the last of the man-eating animals.
>
> The balance of nature was destroyed then, so they had to go on. It was like the machines. They started them—and now they can't stop. (36)

"Seeds of the Dusk" (1938), by another prominent sf pulp writer, Raymond Z. Gallun, is a story of the far future in which humans—the Itorloo—have become not further enlightened, humane, and ecological-minded but "cold, cruel, cunning" (230). Even as the Itorloo plan to invade Venus, a resilient, sentient spore plant has followed the sunlight from a dying Mars to earth where, after a long period of growth and preparation, it begins to repopulate and develop its own "civilization." Possessed of the ability to emit electric shocks, the invading plant develops a plague designed solely to extinguish the Itorloo. The earth's remaining animals—represented by an intelligent crow descendant named Kaw—are happy with the resulting succession of dominance: "If the new masters were not truly benignant, they were indifferent" (253). Gallun's story is a parable of evolution as well as cyclical history. The story concludes by suggesting that a future species—perhaps the distant relatives of the Itorloo—may one day return and restore its dominance.

My survey of early sf with antianthropocentric and ecological themes has merely scratched the surface. Most of these works were written for mass appeal for middle to lowbrow readers, and while some of them, as well as scores of others I have not addressed, will likely be read for a long time, many have fallen into neglect and await reconsideration. (One may argue that there are many works in all popular genres, early sf included, that, for all but a handful of literary and cultural historians, deserve their neglect.) Yet, precisely because of their mass appeal, they reached a readership that might not ordinarily consider the idea that humans are not the center of the universe. Some early sf, no less than the poetry of Jeffers, as well as the later sf works I discuss in the next chapter, work rhetorically, forcing readers to question and perhaps alter their attitudes toward exploiting our home planet as if humans were all that mattered. All of these works function much less prophetically than as warnings about the steep, perhaps fatal, price of humankind's inability to be aware of its membership in its land community.

Notes

1. On the possibility that Poe read Volney, see Kevin J. Hayes, *Edgar Allan Poe in Context*. Cambridge: Cambridge UP, 2013: 57.
2. Bierce's word appears to combine a view of northerners as "smug" with the more telling "mugwump," a bolter from the Republican Party in 1884. His definition in *The Devil's Dictionary* connotes a positive term that has become degraded publicly: "In politics one afflicted with self-respect and addicted to the vice of independence. A term of contempt."
3. Cf. The boar Old Major's anthem in Chap. 1 of *Animal Farm*, regarding the coming happy world without humans: "*Soon or late the day is coming,/Tyrant Man shall be o'erthrown,/And the fruitful fields of England/Shall be trod by beasts alone*" (New York: Everyman's, 1993: 7).

CHAPTER 8

Antianthropocentrism and Science Fiction Part II: After World War II and into the Twenty-First Century

This chapter is a picture of how the critique of anthropocentrism developed in post-World War II science fiction up to the near-present, concluding with a focus on Margaret Atwood's MaddAddam trilogy and Cormac McCarthy's *The Road*. By necessity, my primary texts here must be very limited, overlooking such important sf writers who adopt ecological themes as John Wyndham, Clifford Simak, John Christopher, Marge Piercy, David Brin, Nancy Kress, Greg Bear, and Stephen Baxter, among many others. Nor do I make claims regarding the "most important" works during the period. I can only hope to address a few highlights along with what are, perhaps, more obscure works. Perhaps even more so than earlier sf, postwar sf shows the genre's unique ability for extrapolation, a term explored by well-known sf scholar Darko Suvin, who defines sf as "a literary genre whose necessary and sufficient conditions are the presence and interaction of estrangement and cognition, and whose main formal device is an imaginative framework alternative to the author's empirical environment." In sf, then, the reader recognizes a familiar idea in a new, unfamiliar context, potentially providing readers new ways of looking at the world.

Sf writer Paolo Bacigalupi points to the genre's ability to make abstract, complex, and long-term ideas more visceral: "This gap, between what we flee on the savannah and what might destroy us completely in thirty years, is where I make my writer's home" (Foreword xiii–xiv). Brian Aldiss, as quoted before, famously defines sf as "*Hubris clobbered by nemesis*"; the genre is "*the search for a definition of mankind and his status in the universe which will stand in our advanced but confused state of knowledge (science)*

B.L. Moore, *Ecological Literature and the Critique of Anthropocentrism*, DOI 10.1007/978-3-319-60738-2_8

and is characteristically cast in the Gothic or post-Gothic mode" (25–26; italics in orig.). The search for our "status in universe" is especially tied, at least much of the time, to ecology (and ecological literature), which concerns the interrelations of species. Aldiss's definition highlights sf's unique ability to imagine, to dramatize a world without humans, or a world in which our role is reduced significantly. I hope to show that the critique of anthropocentrism is central not only to ecocriticism but to much postwar sf.

From the 1940s to the Late 1960s

The effects of the two world wars on sf were, of course, momentous. After World War I many were convinced that another, similar war would bring about the end of civilization and age of barbarism. Many sf works reflect this idea, including Walter Van Tilburg Clark's "The Portable Phonograph," which dramatizes the struggles of a small pocket of survivors after a nuclear holocaust, yet the story was published three years *before* atomic bombs were dropped on Japan. This apocalyptic mindset provided fuel for speculation on the desirability of a new race and alternate world possibilities as suggested in Olaf Stapledon's *Last and First Me*n and Arthur C. Clarke's *Childhood's End*. World War II and the new reality of nuclear weaponry raised further fears, but after a while, the atom bomb was largely taken for granted. American writers were less affected by World War I (which played out far from North America) than World War II. In the early pulp era, marked by Hugo Gernsback's introduction of *Amazing Stories* in 1926, few writers and readers were concerned with global catastrophes, but the second world war changed this. More interest in technologies and interplanetary travel developed, and by the time John W. Campbell took over *Astounding Stories* in 1937, pulp sf began to become more sophisticated and intellectual.[1]

George R. Stewart's celebrated 1949 novel *Earth Abides* is infused with a strong ecological viewpoint, and it extrapolates a world in which an anthropocentric mindset is all but impossible; much of the novel's power lies in its dramatization of a world with a much reduced human presence. The book's main character, Ishwerwood ("Ish") Williams—likely a borrowing from accounts of Ishi, the last member of the Yahi tribe of California[2]—has avoided contracting a virus that has wiped out almost all of humanity by his isolated geographical work at a remote post in California and, he reasons, through the chemical reactions of a rattlesnake bite he suffered. He decides to continue his existence, number one, in order "to see what will happen in a world without man, and how" (38). Ish concludes that the desert existed long before the "Great Disaster," as he later calls the catastrophe, and will continue to exist long after man is gone (49). More an ecological book than one plotted in a fantastic sf idiom, a large part of the novel's continued interest lies in its speculation on how the natural world would proceed without humans. Homeostasis alters considerably—and favorably. Cattle disappear, but the wolves and other predators flourish (50). A "peace" settles on the world when man, the "King of Beasts," disappears (117). Yet domestic flora and fauna fade quickly: dogs and cats go wild (27), and grass and flowers wither (43). As nature prevails over human development, it reclaims its full ability to order itself by fires (318).

Like few other works of fiction, *Earth Abides* is filled with images that force characters to reevaluate the place of humans on the earth. Even though Ish is a scientist, he finds some of these realizations jarring. Post-catastrophe, he convalesces from the snakebite and considers how he once imagined that the stars look down on earth. If they did, he concludes that (echoing the conclusion of Wells's "The Star") even though human presence is all but gone from the earth, they "*saw no change . . . Seen even from the moon, the planet that night must have shown only with its accustomed splendor—no brighter, no dimmer*" (17). Later, looking at bridges on the San Francisco Bay, he likens his feelings about them, signs of the glory of a lost civilization, to those of a Burgund or Saxon tribesman looking at a Roman gateway or triumphal arch, but he concludes that the analogy will not work: "The tribesman was sure and content in his own ancient folkways," while Ish realizes he is more like "the last of the old, a surviving Roman" (188–189). Although the struggle for existence remains difficult for Ish and the other survivors, he does not fall into complete despair. His observations of nature's fecundity without humans restore him to some degree, and he settles into thought that resembles closely that of Aldo Leopold's land ethic (as stated in *A Sand County Almanac*, also published in 1949). Coming upon a rattlesnake on a long-deserted college campus, he doesn't see the point of killing it, given the new order. Before the catastrophe, "men had really felt themselves as the master of creation. Everything had been good or bad in relation to man. So you killed rattlesnakes. But now nature had become so overwhelming that any attempt at its control was merely outside anyone's circle of thought. You lived as a part of it, not as its dominating power" (289).

Though written in the context of the mid-twentieth century, the novel subtly subverts consumer society: twenty-one years after the disaster, Ish observes that a grocery store is "depressing," even "horrible" (195). Ish eventually locates other human survivors and becomes the head of a new, local civilization. Children born after the catastrophe refer to older people such as Ish as "the Americans" (228), and over a mere generation, established names and knowledge disappear rapidly from human consciousness (cf. London's *The Scarlet Plague* and Miller's *A Canticle for Leibowitz*). Stewart's novel envisions not so much a world without humans —society will undoubtedly make a comeback in some guise, over time— but in its demonstration that civilization's veneer is thin and fragile, while nature will continue and even flourish without humans: earth abides. It also shows, in line with various works by Byron, Cole, Bryant and others, the transience of what we commonly assume as permanent and how the "ruin of one civilization would pile up on the ruin of another" (54).

Although not known for sf, E.B. White, in his *New Yorker* story "The Morning of the Day They Did It" (1950), attacks the overreliance on technology and, is, in its attack on the "debasement of taste" and the welfare state, comparable to ideas in Evelyn Waugh's more reactionary "Love in the Ruins." Between, and by way of, jokes, though, White is, at mid-century, most concerned with basics—our alteration of the environment, the balance of nature, our food sources, and human hubris. In an unspecified future America, the "it" of the title refers to the cataclysm caused by a mishap with an Army-run station out of the earth's orbit, which sprays streams of the toxic "Tri-D solution," a pesticide which "had revolutionized agriculture, eliminated the bee from nature, and given us fruits and vegetables of undreamed-of perfection but very high toxicity" (134). Birds are also extinct, excepting the whooping crane, though a crane sanctuary becomes ground zero for White's apocalypse. Cows are now fed intravenously and stood up so they might be milked continuously, a technology upon which Melonie Babson (who has just written a bestseller on euthanasia) remarks, "it isn't the hubbub that counts, it's the butterfat" (136).

A Cold War story skeptical toward nationalism, the story considers the feasibility of mutually assured destruction. Predating Reagan and Star Wars (SDI) by decades, the Army establishes a weapons system, including the "New Weapon," which replaces the H-Bomb, in space. Americans are flying high, filled with national pride, but when a lieutenant and major break gravity, they discover that their consciences have disappeared, and they initiate the destruction of the planet—"a mild irony" (148). Ever since birds became extinct the narrator had been "reasonably sure that human beings were on the way out, too" (149). People before the catastrophe, he noticed, had displayed a growing impatience with their surroundings and sought meaning in television, "the universal peepshow": "Only what had been touched with electronics was valid and real. I think the decline in the importance of direct images dated from the year television managed to catch an eclipse of the moon. After that, nobody ever looked at the sky, and it was as though the moon had joined the shabby company of buskers" (150). He relates the story from "an inferior planet" to which he is nonetheless warming up: it has worms in its apples, but he likes them better than the ones on his late home planet.

Long associated with New York City, White was at heart a nature lover. This is evident in "The Morning of the Day They Did It" and in some of his essays, including "Sootfall and Fallout," in which he observes the smog over Turtle Bay in Manhattan in 1958: "I don't know what new gadget the factories of Long Island are making to produce such a foul vapor—probably a new jet applicator for the relief of nasal congestion" (113). White then shifts his critical eye to the presidential election season and tries to place an event laced with such self-importance into a larger context: "The impression one gets from campaign oratory is that the sun revolves around the earth, the earth revolves around the U.S., and the U.S. revolves around whichever city the speaker happens to be in at the moment." White calls this the "un-Copernican system." Our gradual contamination of the planet adds up, he writes, "to a fantasy of such grotesque proportions as to make everything said on the subject seem pale and anemic by contrast. I hold one share in the corporate earth and am uneasy about its management" (115). White would give his vote to the candidate that stands up for mud turtles; such a candidate may have lost his reason, but he will have "kept his head . . . wilder regions of thought where the earth again revolves around the sun" (122). White is, in all but name, attacking our national anthropocentrism, which is perhaps even more pronounced and aggressive now than in late 1950s.

Frederik Pohl's *The Space Merchants* (1953; rev. ed. 2011), co-written with Cyril Kornbluth, burlesques corporate advertising, its cozy relationship with government, and its degrading effects on gullible citizens, but it is also ecologically informed in its realization of how unchecked capitalism can destroy the planet for its resources. Published during the thick of the McCarthy hearings on Un-American activities, the novel's main character, Mitch Courtenay, crack copywriter for an ad agency, is abducted by a group of "Conservationists" or "Consies," a group that shares characteristics with "Commies." One of the novel's Consies tells Courtenay, "think what Venus means to us—an unspoiled planet, all the wealth the race needs, all the fields and food and raw materials" (87). The Earth of the novel is no longer fit for life, and humans are mere exploiters and plunderers. Pohl benefitted readers with a long career. In his 1994 very short story "Creation Myths of the Recently Extinct," planet colonizers ponder different "cleanup" methods for wiping out "a little blue planet" that is "infested" with life. They consider striking it with a giant asteroid, but one of the crew reminds them that when they tried that sixty-five million years before, it only succeeded in wiping out "the big scaly things with the sharp teeth." The crew decides to set down a "Black Monolith" among the planet beings (a nod to Arthur C. Clarke's *2001: A Space Odyssey*) that will teach the primitive creatures to use hands, tools, and eventually lead to overcrowding; after "a million revolutions" of the planet, "those little creatures'll have the planet scorched sterile in no time!" (408).

Few sf writers are more discussed by posthumanist scholars than Philip K. Dick, and many of his works directly and indirectly confront ecological issues. His story "Autofac" (1955) is set on earth in the wake of a war involving hydrogen bombs. Most surviving animals, including most humans, have died out, and the earth is ravaged. A group of saboteurs seek to end the robotic "autofac" system programmed to extract the remaining resources deep under the earth's crust, though their efforts are unsuccessful. Scheming on how to destroy the automatic factory system, they behold the postwar landscape:

> The area was wild. No human settlements existed for miles; the entire region had been seared flat, cauterized by repeated H-bomb blasts. Somewhere in the murky darkness, a sluggish trickle of water made its way among slag and weeds, dripping thickly into what had once been an elaborate labyrinth of sewer mains. The pipes lay cracked and broken, jutting up into the night darkness, overgrown with creeping vegetation. The wind raised clouds of black ash that swirled and danced among the weeds. Once an enormous mutant wren stirred sleepily, pulled its crude protective night coat of rags around it and dozed off. (10–11)

As much as any sf writer, at least in the 1950s, Dick accounts for the whole landscape—weeds, ruins, a mutant wren, later a night moth, stars, the mist, mosquitoes—not just the human, thereby suggesting that the human viewpoint is not the only one that matters. Beyond mere setting, Dick allows us to see in detail his posthuman landscape. Will it be peopled again? The story is pessimistic and holds implications for a world (now more so than 1955) increasingly dominated by automation. We can only hope that Dick's fiction remains mostly fictive. In novels such as *Flow My Tears, The Policeman Said* (1974), he shows how a society with a veneer of jangly commercialism may be run by a police state with death camps. Romantic nature is all but nonexistent in most of Dick's novels and stories.

The work of another postmodern seer, J.G. Ballard's *The Drowned World* (1962), is a conceivably prescient novel in which civilization is deep under water by a "sudden instability" that has caused the sea to rise (21). Global temperatures have risen to levels inhospitable for humans. Even in London, above which most of the novel takes place, daytime temperatures reach well over 100°, and Kerans, the biologist-hero of the novel, was born in the Arctic Circle (now a sub-tropic 85 degrees (20)) and has no knowledge of twentieth-century Europe. The novel is not explicitly political—less than halfway through, its speculative nature largely gives way to an adventure story, as the antagonist Strangman seeks to drain the lagoon and steal the treasures remaining in the flooded London—yet like other postapocalyptic novels, the opening chapters subvert images of the luxurious twentieth-century lifestyle as Kerans lives the parody of the lush life in an abandoned hotel, the lower parts of which are under water. (Ballard uses a similar device in his short story "Cages of Sand," and Kim Stanley Robinson employs a similar setting in his story "Venice Drowned.") The two worlds—the lagoon and the twentieth-century buildings—are permanently "interlock[ed] . . . suspended" (11). Echoing Richard Jefferies, Ballard's London is a rising swamp, a lagoon possessed of "a strange mournful beauty" (10). London is filled with alligators and iguanas, which are "again the dominant form of life" (18).

As the human birthrate declines, flora and fauna assume quickly the forms of the Triassic period (23, 42). Kerans refuses to listen to the news on his radio because "We know all the news for the next three million years" (15). The ruins, which become a routine backdrop for the novel, create "reluctant Venices . . . charm and beauty lay precisely in their emptiness" (21), and such ruins images echo those in romantic poems and paintings I have discussed in previous chapters. People are, in Ballard's world, "like pieces of floatsam" (45). Less successfully, the novel dramatizes how Kerans and others merge psychologically with the new Triassic age (see 71). But the question emerges which hovers over many postapocalyptic works: "was the drowned world . . . an impulse to suicide?" (127).

Ballard has written that his early experiences in a Japanese internment camp in Shanghai, where his father worked while Ballard was a boy, instilled in him the knowledge that humans are not as secure as we like to think we are. His role as a writer, he has said, is to be "a kind of investigator, a scout who is sent on ahead to see if the water is drinkable or not" ("Reality" 4). Ballard often sails into dark waters to explore his antihumanist view of the world. His story "The Drowned Giant" (1964) concisely and unforgettably dramatizes this viewpoint, as the dead body of a giant man with what seem to the sympathetic narrator dignified features of Greek sculpture washes onto a British beach. The locals react initially to the body with uncertainty, then with meek curiosity, and then with more daring, as a few people walk onto the corpse while others clap in celebration of their supposed bold victory. Soon after, more humans, like scavengers on a carcass, venture onto the giant's body and sit on its face "like flies," while others cover its "arms and legs like a dense flock of gulls" on a fish. A gang of youths slides down the face, while others straddle its nose, "and another crawled into one of the nostrils, from which he emitted barking noises like a demented dog" (643–644).

A series of more extreme degradations follow: someone amputates the giant's left hand and head, while others cut swastikas and other signs into its skin. Parts of the giant become material for commodities such as fertilizer, and other parts, including his penis, find their way into the nearby town, the inhabitants of which forget quickly about the wonder of the dead giant. Among other possible readings, the story operates as a parable about both our need for untrammeled wilds and our apparent inability to not deface them. Compared to the giant, the narrator notes that humans are tiny, even insignificant. For him, the giant exists, as does untrammeled wilderness, as a standard by which humans may be defined. The giant "seemed to confirm the identity of my own miniature limbs, but above all, the mere categorical fact of his existence. Whatever else in our lives might be open to doubt, the giant, dead or alive, existed in an absolute sense, providing a glimpse into a world of similar absolutes of which we spectators on the beach were such imperfect and puny copies" (644).

Written in the wake of Carson's *Silent Spring*, Thomas M. Disch's novel *The Genocides* (1965) works from a rather objective, Wellsian point of view to underscore the (seeming) arbitrariness of nature, of evolution.[3] The earth has become a giant farm for unspecified aliens to grow their Plants. The few remaining humans' days are numbered; soon they will be subject to a fiery blaze comparable to the fumigation of a gardener to rid a garden of insect pests. This context provides meaning to seeming throwaway phrases, e.g., the giant Plants are, "like all living things, unwilling to countenance any life but their own" (12). Having displaced humans at the top of nature, "The Plants were alone in these forests, and the feeling of their being set apart, of their belonging to a different order of things was inescapable. It ate at the strongest man's heart" (13). Humans in this inverted order are for safety forced to seek wilderness areas away from urban sprawl (34). If the plants, so aggressively invasive, imperialistic, and unstoppable, are things of nature, can they be bad? What is clear is that they are too large and numerous for Anderson, his sons, and the other survivors to do anything about. As the educated Jeremiah Orville explains to the uneducated, tyrannical, patriarch(al) Anderson, the colder climate persists due to decreased carbon dioxide (a variant on the real-life climate change we now face). Anderson is a scoffer, since this is not explained in the Bible, the only remaining book, though it, too, is extinguished later (57). The humans-as-pests theme becomes more explicit as the survivors, crawling for survival into a cave and then into the root system of the Plants, "were worms, crawling through an apple" (83). Orville and his young lover Blossom ponder resetting humanity as a new Adam and Eve (134), but, escaping through an enormous tuber, Orville and the few survivors behold a scorched, sterile earth in the wake of the alien harvest (143). Amidst the landscape, soon showing clear signs of a second harvest, the remaining humans are "very, very small" (145), and the novel concludes with a quotation from Job 25: compared to the moon and stars, "How much less is man, that is a worm? and the son of man, which is a worm?"

Two of the most significant sf novels with ecological themes published in the mid-1960s are Frank Herbert's *Dune* (1965) and Harry Harrison's *Make Room! Make Room!* (1966). Harrison's novel, as the title indicates, is a strong statement on human overpopulation, a phenomenon that places much stress on natural resources and the quality of all life, including humans. Set in 1999, the world population of the novel is around seven billion people (as prognosticator, Harrison was close, though that number has since then been exceeded). Herbert revolutionized the genre in his intricate ecological description of the desert planet Arrakis, a resource for the precious substance "melange" ("spice"), which can prolong human life and enhance perception, among other benefits, though it is also addictive and dangerous. As Joan Slonczewski and Michael Levy write, "the novel foreshadows the agonizing struggles of the future between near-term utilization and long-term preservation of natural resources" and was tremendously influential for others writing about the ecologies of other planets, including Slonczewski herself and Kim Stanley Robinson (183–184).

Science Fiction and the Age of Ecology

Sf ecological catastrophe literature seems to have peaked in the early 1970s, in the wake of many public environmental disasters and the establishing of the Environmental Protection Agency. This period has been, to date, the most sustained one for environmental reform in American history (Turner 248). Several sf novels and stories appeared around this time, as did anthologies, including *Nightmare Age* (1970), edited by Frederik Pohl, *The Wounded Planet* (1973), edited by Roger Elwood and Virginia Kidd, and *The Ruins of Earth* (1973), edited by Thomas M. Disch. A number of speculative writers active from this period (and in some cases, before) to the near-present reached perhaps just beyond reality to satirize a world of runaway greed filled with pollution and garbage. These writers include Stanislaw Lem, Ursula K. Le Guin, Kurt Vonnegut, Jr., Joanna Russ, James T. Tiptree, Jr., Kate Wilhelm, Edward Abbey, Daniel Quinn, Octavia E. Butler, Kim Stanley Robinson, and Paolo Bacigalupi.

Polish writer Stanislaw Lem's *The Futurological Congress* (1971) is set in an unspecified future characterized by massive overpopulation and worldwide societal breakdown. Similar to Huxley's soma-filled *Brave New World*, technocrats have introduced to the world "pharmacocracy" (97). Reality, as such, is squalid, dehumanized, based on the most ridiculous of profit scams (127–128), while a drug for all purposes gives the people the needed illusion that not only is all okay, but everything is, through technology, the best it ever has been. Everyone is "psychotroped" (127). The protagonist, Ijon Tichy, a recurring Lem character, time-travels to this future world and, by slow revelations, discovers that all is not well. He hears of some who wish to genetically repopulate the earth with animals, which have gone extinct, though (looking back to Wells's *Moreau* and forward to Atwood's MaddAddam trilogy) he believes "we should populate our preserves with bold, original conceptions, not slavish imitations," and he proposes new animals such as "pangaloons" and "luminigriffs" (95–96). Unable to escape a thoroughly chemical-dependent society, a world in which "the distinction between manipulated and natural feelings has ceased to exist," Tichy proposes to his friend Professor Tarantoga that they "hole up somewhere in the Rocky Mountains" (120, 137). Many of

Lem's stories hold strong though also often comic ideas regarding the human impulse to trash their environment, including "Let Us Save the Universe," in which the narrator (again, Tichy) details how "cosmic tourists" have defaced the Solar System and beyond: "Beyond Sirius I began counting the huge signs advertising Mars vodka, Galax brandy, Lunar gin, and Satellite champagne, but soon lost count" (141).

Ursula K. Le Guin published many of her most enduring works during the 1970s. Written from a strongly anthropological perspective (both parents were anthropologists), Le Guin, a longtime resident of the Northwest United States, is also a lover of nature and trees in particular, and this is reflected in many of her works. Her brief novel *The Word for World Is Forest* (1972), one of several of her loosely-connected Hainish novels, is ecologically-oriented from its first page onward. The peaceful planet Athshea, populated by a limited number of small, gentle, sentient beings, is invaded by spaceships filled with colonizing humans (yumens), who, the earth having been deforested (almost the only wild animal left on earth is the rat), begin quickly to chop down all the forests of "New Tahiti," as the Terrans/yumens call their Athshea base. From early on, the invaders are wasteful, arrogant, presumptuous, wholly anthropocentric, and blind with machismo (women have been shipped onto the planet for breeding). Athshea is destroyed utterly by the rapacious Terrans, and bear comparison with Ray Bradbury's *The Martian Chronicles* and Dick's "Survey Team," in which humans have scorched not only the earth but Mars as well. Though the novel is, like many other works published in the 1970s, an ecological allegory, it also operates as a critique on white colonization. It evokes the genocide against American Indians as well as the enslavement of Africans (the diminutive native people—demeaningly called "creechies"—become the yumen slaves, though a quiet rebellion repels the yumens), yet it is difficult to understand the novel, which is characterized by unusually bitter language, outside of its Vietnam War era context. The renegade Captain Davidson, the embodiment of anthropocentric, machoistic, ultra-national imperialism, is determined to drop "jelly bombs" on the peaceful, fleeing natives (153).

Many other Le Guin novels hold ecological implications, including *The Dispossessed*, *The Lathe of Heaven*, and *The Telling*, all of which are set within the context of degraded environments. Her fantasy works, including her Earthsea works, also indicate a strongly Taoist, ecocentric viewpoint, as do many of her short stories. Her 1975 dystopian story "The New Atlantis" is prescient: the land on the earth is sinking, due partly to the greenhouse effect, while fresh terra is rising. Low-lying cities such as New York and Miami are under water, and few can afford to purchase house-boats. The story largely revolves around the possibility of alternative energies that are suppressed by the government. Various anarchists provide the only dwindling hope in the story. "Direction of the Road" is told from the point of view of an oak tree that chronicles its views on changing modes of human transport over almost two centuries on the road next to which it stands. The tree is pleased enough with humans when they walk or travel on horseback, but the automobile, a "wretched little monster" (269), throws perception into a new "Order of Things," under which the tree nonetheless does its duty to "share in supporting the human creatures' illusion that they are 'going somewhere'" (272). But one night a speeding car runs into the tree—from the tree's point of view, it is the tree than runs into the car—and kills the driver, causing the tree to ponder the fact that the driver, for an instant, saw the tree "under the aspect of eternity." This, for the tree, "is unendurable. I cannot uphold such an illusion. If the human creatures will not understand Relativity, very well; but they must understand Relatedness" (273–74).

Over a long career, Kurt Vonnegut, Jr. confronts environmental ideas in some of his novels and stories. The plot of his 1972 story "The Big Space Fuck" (1972) follows the story title rather literally, as a rocket ship with a nose filled with freeze-dried "jizzum" is launched to explode in Andromeda: "In 1979, America staged the Big Space Fuck, which was a serious effort to make sure that human life would continue to exist somewhere in the Universe, since it certainly couldn't continue much longer on Earth. Everything had turned to shit and beer cans and old automobiles and Clorox bottles" (207). In his 1959 novel *The Sirens of Titan*, the whole of human history, it turns out, is the result of the manipulation of history by an alien race, the Tralfamadorians, to deliver a small spare part to a robot. Douglas Adams would employ a similar device in his very popular 1979 novel *A Hitchhiker's Guide to the Galaxy*, in which the earth is destroyed to make way for galactic highway by an alien race called the Vogons.

Ecology is not central in Joanna Russ' sf novel *The Female Man* (1975), but the all-female residents of the future utopia Whileaway choose an agrarian lifestyle, even though they are highly technologically advanced.[4] A plague hit the planet many centuries earlier, wiping out all males; the planet has no cities, and inhabitants prefer walking to driving automobiles (12–14). A strongly feminist, influential time travel work built around the narratives of four female characters (variations on the same woman), one of them, Janet Evason explains time/place travel by stating that "there must be an infinite number of possible universes (such is the fecundity of God) for there is no reason to imagine Nature as prejudiced in favor of human action" (6–7). Yet Russ is much more critical of androcentrism than anthropocentrism.

Like Russ, James Tiptree, Jr. (Alice B. Sheldon) is strongly feminist and ecologically aware in her short stories. In "The Last Flight of Dr. Ain" (1969/1974), the title character rubs out the human race to save the earth from a terrorist-induced plague and ecological disaster. One of Tiptree's most celebrated stories, "The Girl Who Was Plugged In" (1973), is set "not all that far in the future" (44) on an earth inhabited by fifteen billion people. Since advertising has become illegal, it operates through indirection: a beautiful fifteen-year-old girl named Philadelphia Burke, a.k.a. Delphi, in reality a corporate, satellite-controlled "Remote," causes viewers of the holocam, potential consumers, to overlook environmental disaster: "a lot of strip mines and dead fish have been scrubbed, but who cares with Delphi's darling face so visible?" (55). Written long before the Internet and social media, the story also may bring to mind big energy TV commercials proclaiming their love for the environment, but the story is also a strong statement about the exploitation of the female body. In a commercialized, consumerized world not unlike our own, above all, people want "Things" marketed by "goddesses" such as Delphi, by whose time "the hunt for new god-gear is turning the earth and seas inside-out and sending frantic fingers to the stars" (56).

Kate Wilhelm's most celebrated sf novel, *Where Late the Sweet Birds Sang* (1976), is often cited as the most important in the genre regarding human cloning. Set in a rural, postapocalyptic Virginia after a human-induced catastrophe, a large group of people, many of them clones, live on in an isolated community, Wilson Farm. Grandfather Wilson explains the catastrophe by saying that the "earth needs a rest," suggesting that the thinning out of humans is God's intention (33). Since surviving humans are infertile, cloning is the only option for propagating the species. Outside the farm, society is tanking. Wars ravage the Middle East, while bands of marauders and plagues spread across the Mediterranean region (43). Later, Philadelphia is burned, as is Baltimore; New York City is irradiated. The trees are radioactive and the land is dead (204, 207). North America is going into a deep freeze, as a glacier appears in Philadelphia (232). In many postapocalyptic stories, humans go back to a state of nature, but the cloned humans that populate Wilhelm's novel are scared of wild nature. Woods are a "hostile environment" (175), and no one can understand why anyone would want to be alone. One of them, Molly, shudders thinking about spending a night in the woods, yet she finds a sense of independence later on the river, which "seemed to have a voice, and infinite wisdom" (102). When she is separated from the others and lost in the woods, she is exiled as a threat by the other clones, who believe that she has become contaminated.

Where Late the Sweet Birds Sang is not only a novel about cloning or about a postapocalyptic civilization; it also functions as a parable about freedom, living well, and how interacting with wild nature makes us more fully human. Molly's son, Mark, is born of conventional birth and is uniquely able to think for himself. Even more so than Molly, he finds his individuality in the woods. The trees talk to him, a recurring theme in the novel (151, 152, 173–74, 206). Most of the cloned characters remain suspicious of individuals, while a few, such as Mark and Molly, are at home in the woods. Some believe Mark useful in training the others how to live in the woods, but others would like to chop down all the trees. Over a little time, the old knowledge of things disappears, as "children lacked imagination to abstract, to fantasize" (195). Mark leaves the community along with some fertile humans to start over, and they find success and live harmoniously. He returns to the compound twenty years later to find that it has perished, that the cloned humans have disappeared, and that nature has reclaimed the compound. Traveling by boat down the Susquehanna River to Chesapeake Bay, he sees no signs of humans. On the beach of the river he looks into the black forest and, receiving validation for his nonconformist lifestyle, he sees "nowhere another person, nowhere a sign of human life, no buildings, no roads, nothing. Suddenly he threw his head back and laughed, a joyous, almost childlike laugh of triumph" (201–02).

One of the most important and radical voices to address the connection between wilderness and human freedom is Edward Abbey, who published one sf novel, *Good News* (1980). Although it is not his best work, it is an interesting variant on a lifetime devoted to praising the wilds of his beloved desert Southwest and attacking its industrialization. The title of the novel is ironic, since the "good news" is that industrial society has collapsed, the result not of a nuclear holocaust but the implosion of an unbridled industrialization. The American landscape is dominated by corporate machinery, and "the ever-growing cities assumed the shape of a nightmare. Not a nightmare of horror but a nightmare of dreariness, a routing and customary tedium" (1). The United States along with all other nations have lost the ability for self-sufficiency, including basic needs such as food (3). Against those remaining who wish to rebuild the old military-industrial complex, represented by the tyrannical Chief, Abbey's heroes, the old, curmudgeon Jack Burns (the protagonist of his 1956 novel *The Brave Cowboy*) and his American Indian friend Sam Banyaca, team with a ragtag group of anarchists in attempting to keep government decentralized. The novel is the outgrowth of Abbey's long-held wish for the fall of industrialized civilization, which, Abbey writes elsewhere, "is the basis of my inherent optimism"—a return "of a higher civilization: scattered human populations modest in number that live by fishing, hunting, food gathering" (*One Life* 28). Hence, as Sam and Jack ride their horses along an abandoned highway, they see

> derelict automobiles at rest on flattened tires, doors sagging open, mice, moles, birds nesting in the ruin of their interiors. Vines and weeds grow from the rotted upholstery of dead Fords, defunct Chevrolets, moribund power-wagons, decayed Cadillacs, and wasted Winnebagos. Quiet bats flicker in, flicker out through the broken windows of tractor-trailer rigs, four-wheel-drive pickups, Blazers, Broncos, Scouts, Jeeps. One thing—besides futility—these machines have in common: All are facing east, away from the city. (29)

Though not sf, *The Monkey Wrench Gang* (1975), perhaps Abbey's best novel, also employs apocalyptic warnings that recur throughout the author's writings from the 1950s up to his death in 1989. The financing member of the gang of eco-saboteurs, Doc Sarvis, sees hope for the overly industrialized country only in catastrophe (42). Another gang member, Hayduke, has a similar viewpoint: "When the cities are gone, he thought, and all the ruckus has died, when sunflowers push up through the concrete and asphalt of the forgotten interstate freeways . . . then maybe free men and wild women . . . can roam the sagebrush canyonlands in freedom" (107). In a sense, Abbey's writings follow Orwell's dystopia to a logical conclusion: tyrannical governments working in tandem with giant corporations want to make wilderness illegal because it suggests too strongly the possibility of freedom. Abbey's four gang members are aware of an order that is independent of, superior to, and more lasting than industrious humans. In the desert, "each plant is separated from its nearest neighbor by ten feet or more," and gang member Seldom Seen Smith travels down a road, a "path pioneered by deer and bighorn sheep twenty thousand years before" (390). Whereas writers such as Jeffers and Thomas Pynchon (whom Abbey has more than once called his "hero") among other twentieth century writers have concluded that humans are not fit for self-reliance, Abbey maintains a certain idealism: "The horned toads, the hawks, and the coyotes and the rattlesnakes and other innocent creatures I hope will survive and carry on, and yes, probably a few humans with them, or at least I hope so. I think the human race will get one more chance. I'm not sure we deserve it, but I hope we get it anyway" ("Bloomsbury" 154).

Daniel Quinn's *Ishmael: An Adventure of the Mind and Spirit* (1992) is, on its surface, more of a fantasy than sf novel, but it is filled with scientific ideas and is one of the few popular works of fiction to address anthropocentrism directly. Although uneven as work of fiction, Quinn's novel succeeds uniquely as a dialogue on the place of humans in the world between a giant speaking gorilla/teacher named Ishmael and the unnamed student/narrator. Sold to a menagerie from a zoo, Ishmael says that he can't understand why his human spectators do not consider themselves animals (15). Over the course of the novel, the narrator learns slowly from Ishmael, who has a unique point of view as both endangered animal and captive animal. Ishmael tells the narrator that humans are also captives—"of a civilizational system that . . . compels you" to destroy; unlike gorillas, humans are not able to find the bars of the cage. Ishmael divides humans into two categories, the "Takers" (the "civilized" living mostly in the West, including the United States) and the "Leavers" (tribal cultures, including American Indians). Ishmael is nonplussed that people can take note of the ozone layer diminishing with such complete calmness (44).

The license by which the Takers assert their charge to exploit the planet is based on anthropocentrism. Man is the "end" of the world, explicitly or implicitly, and the end of creation/evolution, state the Takers. Establishing the myth of Western society that all exists for humankind, the narrator is forced to acknowledge Ishmael's statement that, in terms of the web of life, "Man's appearance caused no more stir than the appearance of jellyfish" (58). The world, the Takers assume, is one giant "human-life support system"—another myth (59). Takers deny that humans are part of the earth's community of life. Humans are not exempt from other laws—gravity, the second law of thermodynamics, and so on; nor are they exempt from the laws of the community of life that state that man is one of many of the earth's living creatures. It matters little whether humans agree that they are a part of the community of life; it is a fact, and humans are not the biological exception to the laws of life (102). People have begun to accept two of the three "dirty tricks" played on man: the idea that man is not the center of the universe (confirmed by modern science), and the idea that man evolved from the slime. The third "trick" he does not accept is the fact that he is not exempt from natural laws (103), and this is precisely what will lead to his own destruction. But the law is "catching up with them," says Ishmael (108). Alone among the earth's creatures, humans exterminate their competitors, systematically destroy and their competitors' food supply, and deny them access to food (126–27).

Octavia E. Butler's novel *Parable of the Sower* (1993; followed by *Parable of the Talents*) could be the world of McCarthy's *The Road*, a decade or so before its events. Like the father and son of McCarthy's novel, Butler's heroes confirm a stubborn humanity by vowing to not eat human flesh (301). The novel's narrator/diarist, Lauren Olamina, infuses the otherwise grim, violent novel with an abiding humanity. The story begins *in medias res*, as a nondescript catastrophe, or set of them, has taken place —cholera, tornadoes, blizzards, nature out of balance (53–54). The climate has warmed, and the sea level—the novel is set wholly in California— has risen, though Lauren's preacher father, living in the face of the disaster, stubbornly refuses to believe that God would allow people to change the earth's climate (57). Lauren and her family are left in a tenuous community to survive gangs of thieves who take drugs, paint themselves blue, set things on fire, and watch them burn. The relationship between animals and humans has changed: dogs are no longer pets but are feral, things to be feared. The novel is at least as much a comment on recent inner-city politics as it is about the imagined future: Lauren's young brother can't make a living except through gang activities, and the police respond late when at all to emergencies in Lauren's neighborhood. There is no functioning government per se, though some of its structure remains. The economy has also collapsed, and inflation is staggeringly out of control. Only those who are very wealthy and who can afford their own private armies maintain anything like normal lives. Lauren succeeds by grasping and accepting change, which she equates in her new religion (Earthseed) with God.

In many of his works, including his Mars trilogy, Kim Stanley Robinson demonstrates not only a keen sense of science but also an understanding of the often contentious and sordid political relations that determine the treatment of natural environments. His 2004 novel *Forty Signs of Rain*, the first in the Science in the Capital trilogy, shows the interrelationships between political maneuvering in Washington, D.C, filled with lobbyists, and the condition of the earth's fragile, dwindling ozone layer. Charlie Quibler is an advisor to a senator who is relatively aware of the need for sound environmental science as a ground for policy, but even he is far from ideal. He tries to aid representatives of Khemelung, a member of the "League of Drowning Nations," but the best a moderate senator can manage is, "I'll see what I can do!"—a euphemism for "probably not." Charlie meets with the president's science advisor, Dr. Zacharius Strengloft (a perhaps too-obvious play on Kubrick's Dr. Strangelove), who cools Charlie's wish to move to combat global warming with the reply that is it is easier to destroy the world than alter capitalism (156). Another central character, National Science Foundation (NSF) microbiologist Frank Vanderwal, roundly criticizes the misplaced interests of the NSF and its "ridiculously passive position" toward global climate change (210). Like Charlie, he finds that economic interests in the United States government trump everything else. When a chunk of ice the size of France breaks off from Antarctica, he notes that it hardly registers in the news (236). The NSF, says Frank, is underfunded to keep this world of "dying dinosaurs" alive (319). When, after 300+ pages of mind-numbingly frustrating politics, a hyperniño hits the West coast, some are surprised, having become jaded by predictions of perfect storms due to global climate change. The flood causes severe mudslides in San Diego, while in Washington, which has an elevation of ten feet above sea level, the National Mall quickly becomes a lake. Frank, stranded in a building by the flood, observes the War College Building "sticking out of the water like a temple of Atlantis" (371), while the Washington Monument is a "dim obelisk on a watery horizon" (374). Although "the capital of the hyperpower" is in hours "drowned and smashed," the news media, driven by profits, continue to show the catastrophic event as a spectacle (378).

Recalling Philip K. Dick's "Autofac," Paolo Bacigalupi's "The People of Sand and Slag" envisions a far-future posthuman and postnatural world controlled by a large corporation, SesCo Mining, which owns and controls mining robots and satellites. Anything found alive on the landscape is considered an "intruder" and a threat to the profits of the corporation. The story focuses on the relationship of a SesCo "security unit," comprised of Chen, Jaak, and Lisa, all of whom are "bio-jobs," enhanced and tailored to corporate needs with mutated bodies that are virtually indestructible. The land of the story is sterile and inorganic, consisting largely of sand, which humans may eat, thanks to genetic modifications, and slag, metal heaps piled over time on the landscape. Into this world appears a dog, a species that is all but extinct. After considering whether to "slag" it, Jaak decides to keep and care for it until he realizes it does not fit well into his world and lifestyle. The story explores the need of humans for fellow animals to help define us, though it concludes pessimistically that "people" in such a cold, sterile future may genetically lose the need for other living beings: "Who needs animals if you can eat stone?" Jaak asks (58). The unit, after all, has its console games (Immersive Response—IR) for entertainment, even though a faint, lingering sense of the loss of dog lingers (67). Bacigalupi is one of the bright lights of ecological-minded sf with his novels *The Windup Girl* (2009) and *The Water Knife* (2015).

Atwood and McCarthy

I conclude by focusing on two of the most important twentieth-century and early twenty-first century North American novelists, Margaret Atwood and Cormac McCarthy, both of whom published popular and well-received apocalyptic novels in the 2000s. These two writers are vastly different in both technique and content, and they dramatize very different worlds. While Atwood, in *Oryx and Crake* (2003) and the two other MaddAddam novels, blends realism and satire in her broad cultural critique, McCarthy's spare, stark prose contributes to his decidedly pessimistic view, even as *The Road* (2006) is in respects affirmative. For all their differences, both writers employ considerable scientific knowledge in their fiction; both dramatize ecological catastrophes that are the results of cupidity and what appears to be a human death wish, and both have written fiction grounded in the understanding that humanity is likely a temporary species on the planet. McCarthy and Atwood make the reader think about the place of humans in the world and our extreme anthropocentrism. For both writers, the degradation of culture and the death of nature are finally inseparable. Presently, I draw from novels by both writers but focus on *Oryx and Crake* and *The Road*, to show how ecological and cultural sickness—the result of extreme anthropocentric hubris—functions in the artistic visions of these writers.

Ecological disaster in Atwood's *Oryx and Crake* is the result of a deadly airborne virus created and spread by a giant corporation with the Orwellian name HelthWeyser. The kitschy names for corporations and consumer products prevalent in the MaddAddam novels may cause one to think not only of Orwell, but Emerson, who writes in *Nature*, "The corruption of man is followed by the corruption of language" (14–15). Post-disaster, all traces of human habitation are disappearing rapidly. In the wake of the airborne, corporate-engineered virus, plagues, famines, and floods ravage the earth. The main character of *Oryx*, Jimmy/Snowman, immune from the virus, is one of the few human survivors; his father had worked for "OrganIncFarms" as "genographer" for the "pigoon" project, which entails the Moreau-like harvesting of transgenic pigs for replacing human organs. Engineered "wolvogs" have devoured domestic dogs (108); the bobkitten was introduced to eliminate feral cats and improve the songbird population (164), and the food chain is shaken up utterly (158–59). Post-catastrophe, the human species devolves: Jimmy, feeling a growing need to swing from limb to limb (108), lives in a tree, and he feels the urge to cool himself like a dog (39). Surviving men mark territory with their urine (154), and females, baboon-like, find themselves going into heat on occasion (see 164–65).

Taken together, in the MaddAddam (the name of the genetics complex that created the catastrophic virus through the "Paradice Project") novels, Atwood places a deft finger on our own culture, dominated by fundamentalist ideologies, almost complete corporatism, and brazen, vulgar, and testosterone-driven consumerism. Like most modern-day satirists, Atwood has to sail into some fairly remote waters to go beyond a consumerist world that already accepts absurdity as normalcy, and she pulls it off as well as any Western writer today. The MaddAddam novels reflect thoroughly what Walter Benjamin called "the phony smell of a commodity" ("The Work" 231), a smell deposited in virtually every corner of Atwood's near-future world. As in what is perhaps Atwood's most famous work, *The Handmaid's Tale*, the MaddAddam patriarchal order isn't a side issue but central—the cause—in the degradation of culture and of nature. The source of the catastrophic virus is the Viagra-like product "BlyssPlus," designed by Jimmy's boy-genius friend Crake, who grew up to be a brilliant but nihilistic scientist. Nuclear waste and other industrial poisoning led to the low birth rate central to the plot in *The Handmaid's Tale*, the dystopian world of which (Gilead) combines the worst elements of medieval Christianity with KGB-like enforcement (see 304–05). The world of the MaddAddam novels, which also include *The Year of the Flood* (2009) and *MaddAddam* (2013), events in which are largely contemporaneous, is no less patriarchic. Where sexuality in Gilead is cloaked in illegal backrooms, the pre-apocalyptic world of MaddAddam is brazenly, gaudily sexual, and one may perceive more than a little of this world today, e.g., in some TV beer and erectile dysfunction pill commercials.

Atwood's picture of a degraded and degrading online world reflects the need for more extreme sensation, which results in the debasing of culture and a prelude to the end of the world (as we know it)—"Brainfrizz" (62). Reflecting on how the world had ended up as it is, Jimmy recalls how everything in the run-up to events triggering catastrophe had been "nothing but sublimation, according to the body. Why not cut to the chase?" (85). In the world just before catastrophe, even the veneer of sublimated wholesomeness was removed. The character of Oryx is essentially a sex slave with whom both Crake and Jimmy sleep after initially observing online. In the MaddAddam world, as in our world, to quote Andrea Dworkin, "Capitalism is not wicked or cruel when the commodity is the whore; profit is not wicked or cruel when the alienated worker is a female piece of meat . . ." (qtd. in Hedges 55). Even as consumerist-corporatist culture has gone over the top, a ravaged earth is also in a tailspin, a contemporaneous and kindred event. Well before the release of the virus, the earth is in trouble due mostly to global climate change, though the novel does not discuss this directly. At the beginning of the novel, Jimmy, from his tree in the woods, recalls how in his childhood leaves still changed colors. In a flashback (the novel, like *The Year of the Flood* and *MaddAddam*, toggles between a pre- and postapocalyptic world), Jimmy and Crake graduate from HelthWyzer High in early February, now a warm, tornadic season. June has become a wet season on the east coast, and Harvard is now under water (173). Texas has dried up and blown away (244). After the release of the virus, which crosses the globe quickly, all traces of human habitation begin to disappear (see 324). After the virus—the prophesized "waterless flood" in *The Year of the Flood*—Jimmy walks in a once-populated area and notes, "It won't be long before all visible traces of human habitation will be gone," though he has no way of knowing how many people are left (222). Near the end of the novel, he, living among the virus-resistant, engineered humans, Crakers, watches the world from the safety of the high-tech bubble dome: "the end of a species was taking place before his very eyes. Kingdom, Phylum, Class, Order, Family, Genus, Species. How many legs does it have? *Homo sapiens sapiens*, joining the polar bear, the beluga whale, the onager, the burrowing owl, the long, long list" (344). A key to the novel's title, several species of both genera—Oryx and Crake—have gone extinct or are threatened seriously.

The novel *MaddAddam* focuses largely on the character of Zeb, who is a minor character in *The Year of the Flood*. With a deft understanding of cultural nuances, Atwood shows the connection between the fall of culture and destruction of nature—followed by nature batting last and "encroaching" onto human civilization—is explicit in the popular entertainment of the day:

> Speculations about what the world would be like after human control of it ended had been—long ago, briefly—a queasy form of popular entertainment. There had even been online TV shows about it: computer-generated landscape pictures with deer grazing in Times Square, serves-us-right finger-wagging, earnest experts lecturing about all the wrong turns taken by the human race.

But ratings for such entertainment faded, since there was "only so much of that people could stand." The public opts instead for hotdog-eating contests and ultra-violent spectacles, entertainment "so much more palatable than the truth" (32). Later in the novel, Zeb briefly plays a "biogeek challenge game called Extinction" his brother Adam designed in which the player guesses the names of extinct organisms, though the game is "a certified yawner." Zeb speculates that Adam may have designed the game with contempt for the views of their reverend father, whose interpretation of Genesis "was that God had made the animals for the sole pleasure and use of man, and you could therefore exterminate them at whim" (194).

As in Atwood's trilogy, humanity is on the wane and devolving in McCarthy's *The Road*, which is set almost entirely in the wake of the apocalypse, though the emphasis is less on the ravaged environment than on a father's determination to give his young son the best possible chance to survive a very dark time by traveling to the south coast. Everything he thought he knew, the man thinks at several points in the novel, is a "lie" and "[m]ore fragile than he would have thought. How much was gone already? The sacred idiom shorn of its referents and so of its reality" (89). There cannot be a nuclear disaster that is not also an ecological one. In McCarthy's world, nature has fallen into almost complete disorder, and so "the banished sun circles the earth like a grieving mother with a lamp" (32). Though not stated directly, such banishing—i.e., "compulsory removal"—suggests that the catastrophe is completely anthropogenic: humanity is the agent, the banisher. The personified, "grieving mother" sun is no longer able to care for its progeny, life on earth. The earth is "intestate" (130), having died without leaving a will. In his pared-down style, which matches the novel's stark content ideally, McCarthy suggests that the earth has died, leaving humanity without the means to care for itself; the remaining humans in McCarthy's world are, in effect, starving, motherless children.

As the father and son travel down the road, they note the fallen and still-falling trees; even the kudzu is dead (177). Since sunlight is reduced greatly, fish are extinct and all crops have failed (23). Birds are extinct: a man wears a tattoo of a bird, drawn by one apparently ignorant about what birds look like (63). The boy stares for a long time at a mounted deer head in a ruined grocery store (214). The nightmare of a world without fellow animals, which help define us as a species, has been realized. As in Atwood's postapocalyptic world, the more primitive roots of humanity emerge, as humans devolve to more primitive, animalistic states. Theriomorphism—the figure by which humans are given animal qualities—runs throughout the novel. The man and his son confront a member of a "bloodcult"—ragged, cannibalistic, roving bands of which dot the landscape. The man is forced to shoot him when he seizes the boy, and afterwards the man notes the "reptilian calculations," his "cold and shifting eyes" (75). The man and his son, now a part of the radically thinned food chain, also live more like wild animals, more cautiously as, for example, when the man teaches the boy to lie like a fawn in the woods (118).

As with Atwood, McCarthy offers a biting critique of a species that would bring ruin on itself, although his critique is less direct than Atwood's. McCarthy has been attacked by prominent critics, most publicly, perhaps, by James Wood, who in a 2005 article in *The New Yorker* credits McCarthy as "a wonderfully delicate noticer of nature" (qtd. in Hage) but dislikes what he sees as "rhetorical theatrics, male clans, rigged plots, and *The Road*'s 'metaphysical cheapness with a slickness unto death all its own'" (qtd. in Lincoln 10). But in his way, McCarthy is just as critical of consumerism and environmental degradation as is Atwood. Much of the novel is set on a road (the road we are, in a broader sense, going down) on which a man and his son push a shopping cart—a parody of consumer world now in ashes. Ironically, as the man and the boy scavenge for supplies inside a ruined supermarket (a grotesque name, in context), they leave the cart in the parking lot (22). Billboards formerly advertising consumer products are used for crude, frantic messages. They are "whited out with thin coats of paint in order to write on them and through the paint could be seen a pale palimpsest of advertisements for goods which no longer existed" (127–28). Consumerism, thin by its nature, is the first thing to go in McCarthy's novel. The man imagines that cities will soon be "held by cores of blackened looters . . . carrying charred and anonymous tins of food in nylon nets like shoppers in the commissaries of hell" (181).

Suggesting that the veneer of our system of supply and demand is skin deep as well as the truism that no civilization is immune from immolation, near the first of the novel, at the supermarket soft drink machines, the man observes now useless "[c]oins everywhere in the ash" (23). Corresponding to this image, later, in a field he finds an arrowhead, which he presents to the boy, and Spanish coins, remnants of two disparate cultures long vanished (203–04). The earth is now "one vast sepulcher" (222), an old literary metaphor we have seen in Bryant, among others. The boy wears a pinstripe coat that is much too large for him (101)—another subtle indication that human culture is flimsy: a suit coat once typically worn by businessmen and political leaders at the top of the ladder now keeps a starving boy on a dying earth from freezing. They observe a barn upon which is a painted advertisement in fading white letters: "See Rock City" (21). Such ads are common, familiar to anyone who has driven on the highways of the American southeast. (Rock City, on Georgia's Lookout Mountain south of Chattanooga, Tennessee, enables viewers to see seven different states.) The notion of "Rock City" here is ironic, since all cities now are nothing more than rock and rubble, and there are no more states (43). Driving in the south, one also sees many painted signs heralding the imminent apocalypse as foretold in the Bible; here, McCarthy suggests that the ubiquitous "See Rock City" signs are the more credible harbingers of the future.

The world of the novel is, like that in other McCarthy works, cold, dark, violent, and godless, yet an essential human dignity appears to triumph at the end of this, the bleakest of his novels. People are capable of acting heroically. The (apparent) nuclear disaster occurred several years before the novel is set. The boy and man have lasted this long, and it is not clear that humanity will vanish completely. The man and the boy are "keepers of the fire," self-proclaimed "good guys" who practice human dignity: they do not eat people. Ashley Kunsa, citing reviews by Michael Chabon and William Kennedy, argues for an even brighter interpretation of the novel: the boy, born before and thus innocent of a world that would make such a hell on earth, "serves as an Adamic figure, a messiah not unlike Christ himself" posited to remake the world (65). McCarthy told Oprah Winfrey in 2007 that he wrote the novel as a "love story" for his young son (qtd. in Lincoln 164), and this surely accounts for the relative optimism of the novel—its ending in particular. While McCarthy's artistic view probably shares more ground with Samuel Beckett than Oprah, *The Road* is a work of love nonetheless.

In no other work does McCarthy express his view on the fragility of civilization so explicitly as in *The Road*, but it is a recurring, typically subtle idea throughout his writings. In his second novel, *Outer Dark* (1968), one of the main characters, Holme, wanders around the landscape of east Appalachia in search of his sister, who in turn is searching the land for their inbred infant son. Holme finds his way in the dark (variations of which function as a motif throughout the novel) to an abandoned town, Preston Flats, which "looked not only uninhabited but deserted, as if plague had swept and decimated it. He stood in the center of the square where the tracks of commerce lay fossilized in dried mud all about him, turning, an amphitheatrical figure in what moonwrought waste manacled to a shadow that struggled grossly in the dust" (131). His next novel, *Child of God* (1973), plants ideas regarding the smallness of humans against time and forces of nature, as when the strange, violent main character Lester Ballard walks through woods in East Tennessee with "toppled monoliths among the trees and vines like traces of an older race of man" (25). Later in the novel, the town of Sevierville is covered in water as if to suggest Noah's flood (160, etc.). Though Ballard is, as the omniscient narrator says, "A child of God much like yourself perhaps" (4), he is a disturbing but fascinating picture of humanity at its worst or, perhaps, in its brutish essence.

In his nightmarish novel *Blood Meridian*, Judge Holden, the darkest of characters in a novel filled with dark characters, epitomizes an utterly anthropocentric view of the universe. He despises the autonomy of nonhuman nature, and this twisted pseudo-Enlightenment philosophy permits all manner of bloodshed and exploitation (see 198–99). In *No Country for Old Men*, seemingly throwaway details establish the reality that another race of people had inhabited the southwest long ago. The novel suggests that we are headed toward a fate similar to those of Paleo-Indians and later peoples of the southwest. Llewellyn Moss hunts in the desert, where rocks "were etched with pictographs perhaps a thousand years old. The men who drew them hunters like himself. Of them there was no other trace" (11). In the final section of the novel, the novel's first person narrator, Sheriff Bell, recounts a dream: he is riding a horse in the cold with his father nearby: "And in the dream I knew that he was goin on ahead and that he was fixin to make a fire somewhere out there in all that dark and all that cold and I knew that whenever I go there he would be there. And then I woke up"

(309). These final sentences in the novel could serve as a sort of prelude for *The Road*, a novel about a father and son struggling to carry light through a darkened world. Significantly, *No Country* is set in 1980. The sheriff notes that civilization's dissolution due to greed and violence is imminent. "It won't be long, neither," he says (138). Though unstated, the events in *The Road* may take place a generation or so later, in the early twenty-first century. Similar ideas echo also in McCarthy's 2006 play *The Sunset Limited*. One of the two characters, White, says, "The World is largely gone. Soon it will be wholly gone" (25). Near the end of the play, having found nothing to affirm in the earnest but untenable Christian viewpoint of Black, White says, "The darker picture is always the correct one . . . we imagine that the future will somehow be different . . . we will not be here much longer" (112).

A writer, according to J.G. Ballard, is an "investigator and a sort of early warning system." The "cosy suburbia" in which most of us in the west live doesn't prepare one for climate disasters. "Nothing is as secure as we like to think it is" ("Reality" 4–5). Atwood and McCarthy are among the writers who fill this investigative role for us today: the MaddAddam novels, rich ecological knowledge, trace the degradation of culture and nature to a common arrogance, while the starker, realistic tone of *The Road*, less directly ecological but no less grounded in science, forces readers to confront a potential nightmare of our own making. The degree to which Atwood and McCarthy see their fiction as predictive is unclear. In a 2011 NPR interview, McCarthy said, "I'm pessimistic about a lot of things . . . but it's no reason to be miserable about it." We are bad at prognostication, he said; "The fact that I take a rather dreary view of the future is cheering because the chances are that I am wrong." Atwood writes that *The Year of the Flood* is a work of fiction, "but the general tendencies and many of the details in it are alarmingly close to fact" (Acknowledgements).

What is clear is the ability of these novels along with many other literary works, some of which I have discussed in this book, to provide an understanding of where we are positioned in our moment in time, and some of them explore the consequences of ignoring our membership in the land community. The knowledge that we are not the only species that matters should cause us to think and act more responsibly toward our home planet and our fellow species. To ignore that knowledge is to hasten our own dissolution; to begin to more fully accept it would result in a healthier planet. Yet anthropocentrism remains an underdiscussed idea in academia, never mind in the offices of political and corporate power. Our worldview, Loren Eiseley writes, "is still Ptolemaic, though the sun is no longer believed to revolve around the earth" (57). If the understanding and confrontation of anthropocentrism seems to many people an obscure idea and to others a preposterous one, perhaps it is because the idea is too close to home and is unsettling to assumptions about our own significance. We are, after all, only one among millions of species on a planet suspended in an inconceivably massive, expanding universe billions of years old. Ecological science, as well as biology and astrophysics, among other sciences, have demonstrated that, outside of our own domain, the belief in the centrality of our species is a false position. As we have seen, a great many writers since antiquity, imaginatively exploring the implications of science, have already provided parts of the artistic, philosophical, and, in some cases, ecological understanding requisite for lowering the human throne, even if slightly.

Notes

1. Information gleaned from J.A. Cuddon's entry for "Science Fiction" in *The Penguin Dictionary of Literary Terms and Literary Theory*. 4th ed. London: Penguin, 1999 (791–800) is central in this paragraph.
2. The foremost documentarian on Ishi was Theodora Kracaw Kroeber Quinn, Karl Kroeber and Ursula K. Le Guin's mother, who wrote three books on him.
3. Disch makes direct and indirect reference to Carson—see 117.
4. The attitude of Whileaway inhabitants toward nature and technology echo Heidegger on enframing: "We can use technical devices, and yet with the proper use also keep ourselves so free of them, that we may let go of them at any time" (*Discourse on Thinking*. Trans. John M. Anderson and E. Hans Freund. New York: Harper, 1966: 54).

Works Cited

A.I. Artificial Intelligence. Dir. Steven Spielberg. DVD. Warner Bros, 2001.

Abbey, Edward. *Beyond the Wall: Essays from the Outside*. New York: Holt, 1984.

———. "The Bloomsbury Review Interview." By Dave Solheim and Rob Levin. *Resist Much, Obey Little: Remembering Ed Abbey*. Ed. James R. Hepworth and Gregory McNamee. San Francisco: Sierra, 1996. 138–55.

———. *Desert Solitaire: A Season in the Wilderness*. New York: Ballantine, 1968.

———. *Good News*. New York: Dutton, 1980.

———. *The Monkey Wrench Gang*. New York: Perennial, 1975.

———. *One Life at a Time, Please*. New York: Holt, 1988.

Ackroyd, Peter. *Shakespeare: The Biography*. New York: Anchor, 2005.

Adams, Henry. *The Education of Henry Adams*. Oxford: Oxford UP, 1999.

Aldiss, Brian W. *Trillion Year Spree: The History of Science Fiction*. New York: Atheneum, 1986.

Alter, Robert, trans. *The Wisdom Books: Job, Proverbs, and Ecclesiastes: A Translation with Commentary*. New York: Norton, 2010.

Althusser, Louis. "Marxism and Humanism." In Badmington, *Posthumanism*, 30–33.

Altick, Richard D. *Victorian People and Ideas: A Companion for the Modern Reader of Victorian Literature*. New York: Norton, 1973.

Amis, Kingsley. *New Maps of Hell: A Survey of Science Fiction*. New York: Harbrace, 1960.

Ammons, A.R. *The Selected Poems: Expanded Edition*. New York: Norton, 1987.

———. *Sphere: The Form of a Motion*. New York: Norton, 1973.

———. *Garbage*. New York: Norton, 1993.

Aristotle. *On Rhetoric: A Theory of Civic Discourse*. 2nd. ed. Trans. George A. Kennedy. New York: Oxford UP, 2006.

———. *On the Heavens*. Trans. W.K.C. Guthrie. Cambridge: Harvard UP, 1939.

———. *Physics*. Trans. Robin Waterfield. Oxford: Oxford UP, 2008.

Arnold, Matthew. "Dover Beach." *Poetry and Criticism*. Boston: Houghton, 1961. 161–62.

B.L. Moore, *Ecological Literature and the Critique of Anthropocentrism*, DOI 10.1007/978-3-319-60738-2

Atlas, James. "Is This the End?" *The New York Times*. www.nytimes.com. 24 Nov. 2012. Accessed 24 Nov. 2012.

Atwood, Margaret. *The Handmaid's Tale*. New York: Anchor, 1986.

———. *MaddAddam*. New York: Anchor, 2013.

———. *Oryx and Crake*. New York: Nan A. Talese, 2003.

———. *The Year of the Flood*. New York: Anchor, 2009.

Augustine, St. *The Confessions*. 2nd ed. Trans. Maria Boulding. Hyde Park: New City, 1997.

Bacigalupi, Paolo. Foreword. *Loosed upon the World: The Saga Anthology of Climate Fiction*. Ed. John Joseph Adams. London: Saga, 2015. xiii–xvii.

———. "The People of Sand and Slag." *Pump Six and Other Stories*. San Francisco: Night Shade, 2008. 49–67.

Bacon, Francis. *The New Organon*. Ed. Fulton H. Anderson. Indianapolis: Library of Liberal Arts, 1960.

Badmington, Neil, ed. *Posthumanism*. New York: Palgrave, 2000.

———. "Posthumanism." *The Routledge Companion to Literature and Science*. New York: Routledge, 2011. 374–84.

Baigell, Matthew. *Thomas Cole*. New York: Watson-Guptill, 1981.

Ballard, J.G. "The Drowned Giant." *The Complete Stories*. New York: Norton, 2009. 641–49.

———. *The Drowned World*. London: Harper, 1962.

———. "Reality Is a Stage Set: Travis Elborough Talks to J.G. Ballard." 2006. *The Drowned World*. London: Harper, 1962. 1–16.

Barbauld, Anna Letitia. "The Caterpillar." *The Norton Anthology of English Literature: The Romantic Period Vol. D*. Ed. Stephen Greenblatt. New York: Norton, 2012. 52–53.

———. "The Mouse's Petition." *The Norton Anthology of English Literature: The Romantic Period Vol. D*. Ed. Stephen Greenblatt. New York: Norton, 2012. 40–41.

Barnes, Jonathan, trans. *Early Greek Philosophy*. London: Penguin, 2002.

Bass, Rick. *The Ninemile Wolves*. New York: Ballantine, 1992.

Baudrillard, Jean. "Prophylaxis and Virulence." In Badmington, *Posthumanism*, 34–41.

Becker, Carl. *The Heavenly City of the Eighteenth-Century Philosophers*. New Haven: Yale UP, 2003.

Bedell, Rebecca. *The Anatomy of Nature: Geology and American Landscape Painting, 1825–1875*. Princeton: Princeton UP, 2002.

Benjamin, Walter. *The Origin of German Tragic Drama*. Trans. John Osborne. London: Verso, 2009.

———. "The Work of Art in the Age of Mechanical Reproduction." *Illuminations: Essays and Reflections*. Ed. Hannah Arendt. Trans. Harry Zohn. New York: Shocken, 1968. 217–51.

Beresford, J.D. "A Negligible Experiment." *The Road to SF Volume 5: The British Way*. Ed. James Gunn. Clarkston: White Wolf, 1998. 149–53.

Berkeley, George. "Verses on the Prospect of Planting Arts and Learning in America." http://www.bartleby.com. Accessed 2 January 2017.

Berlin, Isaiah. *The Roots of Romanticism*. Princeton: Princeton UP, 2001.

Berry, Wendell. "God and Country." *What Are People For?* Berkeley: Counterpoint, 1990. 95–102.

———. *New Collected Poems*. Berkeley: Counterpoint, 2013.

———. "A Practical Harmony." *What Are People For?* Berkeley: Counterpoint, 1990. 103–08.

———. *A Timbered Choir: The Sabbath Poems 1979–1997*. Berkeley: Counterpoint, 1998.

The Bible. Authorized King James Version. Oxford: Oxford UP, 1998.

Bierce, Ambrose. "For the Akhoond." *Scientific Romance: An International Anthology of Pioneering Science Fiction*. Ed. Brian Stableford. Dover, 2017. 172–79.

———. "A Vision of Doom." *Collected Works: Vol. 5: Black Beetles in Amber*. New York: Gordian, 1911. 46–49.

Bitzer, Lloyd. "The Rhetorical Situation." *Philosophy and Rhetoric* 1 (1968): 1–14.

Boccaccio, Giovanni. *Theseid of the Nuptials of Emilia*. Trans. Vincenzo Traversa. New York: Peter Lang, 2002.

Boethius. *The Consolation of Philosophy*. Trans. Richard Green. Indianapolis: Library of Liberal Arts. 1962.

Bonaventura, St. *The Mind's Road to God*. Trans. George Boas. https://www.ewtn.com/library. Accessed 10 July 2016.

Booth, Wayne C. *The Rhetoric of Rhetoric: The Quest for Effective Communication*. Malden: Blackwell, 2003.

Bostrom, Nick. "A History of Transhumanist Thought." 2005. http://www.nickbostrom.com. Accessed 20 Aug. 2016.

Bradford, William. *Of Plymouth Plantation: 1620–1647*. Ed. Samuel Eliot Morrison. New York: Knopf, 1952.

Bradley, S.A.J, ed. and trans. *Anglo-Saxon Poetry*. London: Everyman, 1982.

Bradstreet, Anne. "Contemplations." *Works*. Ed. Jeannine Hensley. Cambridge: Belknap, 1967. 204–14.

Braidotti, Rosi. *The Posthuman*. Cambridge: Polity, 2013.

Branch, Michael P. "Before Nature Writing: Discourses of Colonial American Natural History." *Beyond Nature Writing: Expanding the Boundaries of Ecocriticism*. Ed. Karla Armbruster and Kathleen R. Wallace. Charlottesville: UP of Virginia, 2001. 91–107.

Breckman, Warren. *European Romanticism: A Brief History with Documents*. Boston: Bedford, 2007.

Brin, David. "How Will the World End?" Salon.com. 17 June 2012. Accessed 18 June 2012.

Brown, Charles Henry. *William Cullen Bryant*. New York: Scribner, 1971.

Brown, Jane K. "Faust." *The Cambridge Companion to Goethe*. Cambridge: Cambridge UP, 2002. 84–100.

Browning, Robert. "Love among the Ruins." *Poetry*. 2nd ed. Ed. James F. Loucks and Andrew M. Stauffer. New York: Norton, 2007. 137–39.

Bruchac, Joseph. "The Circle Is the Way to See." *The Norton Book of Nature Writing*. Ed. Robert Finch and John Elder. New York: Norton, 2002. 811–18.

———. *Our Stories Remember: American Indian History, Culture, and Values through Storytelling*. Golden: Fulcrum, 2003.

———. "Prayer." http://www.babelmatrix.org. Accessed 10 Dec. 2016.

Bryant, William Cullen. *Representative Selections*. Ed. Tremaine McDowell. New York: American Book Co., 1935.

Buell, Lawrence. *The Environmental Imagination: Thoreau, Nature Writing, and the Formation of American Culture*. Cambridge: Harvard UP, 1995.

Burke, Kenneth. *A Rhetoric of Motives*. Berkeley: U of California P, 1969.

Burnet, Thomas. *The Sacred Theory of the Earth*. https://books.google.com/books . Accessed 9 Dec. 2016.

Burns, Robert. "To a Mouse." *Selected Poems and Songs*. Ed. Robert P. Irvine. Oxford: Oxford UP, 2013. 71–73.

Burtt, E.A. *The Metaphysical Foundations of Modern Science*. New York: Doubleday, 1954.

Bush, Douglas. *Science and English Poetry: A Historical Sketch 1590–1950*. London: Oxford UP, 1950.

Butler, Octavia E. *Parable of the Sower*. New York: Grand Central, 2000.

Byron, Lord. *The Major Works*. Ed. Jerome J. McGann. Oxford: Oxford UP, 1986.

Calasso, Roberto. *La Folie Baudelaire*. Trans. Alastair McEwen. New York: Farrar, 2012.

Callus, Ivan, and Stephan Herbrechter. "Posthumanism." *The Routledge Companion to Critical and Cultural Theory*. 2nd ed. New York: Routledge, 2013. 144–53.

Campbell, John W. "Twilight." *The Best of John W. Campbell*. Ed. Lester del Rey. Garden City: Nelson Doubleday, 1976. 24–43.

Campbell, Thomas. *The Complete Poetical Works of Thomas Campbell*. New York: Haskell, 1968.

Campion, Thomas. "What If a Day." *Works*. Ed. Walter R. Davis. New York: Doubleday, 1967. 473.

Capps, Walter H. *Seeing with a Native Eye: Contributions to the Study of Native American Religion*. San Francisco: Harper, 1976

Casale, Ottavio M. "Edgar Allan Poe." *The Transcendentalists: A Review of Research and Criticism*. Ed. Joel Myerson. New York: MLA, 1984. 362–71.

Cassirer, Ernst. *The Philosophy of the Enlightenment.* Trans. Fritz C.A. Koelln and James P. Pettegrove. Princeton: Princeton UP, 1951.

Castiglione, Baldassare. "Superbi colli, e voi sacre ruin." Trans. Mick Swithinbank. 28 Aug. 2016. www3.cpdl.org/wiki. Accessed 5 Sep. 2016.

Catechism of the Catholic Church. http://ccc.usccb.org/. 2nd. ed. Accessed 10 Dec. 2016.

Chandler, James. "Percy Bysshe Shelley." *The Cambridge Companion to English Poets.* Ed. Claude Rawson. Cambridge: Cambridge UP, 2011. 344–59.

Chateaubriand, François-René, vicomte de. *The Genius of Christianity; or the Spirit and Beauty of the Christian Religion.* Trans. Charles I. White. New York: Howard Fertig, 1976.

Chaucer, Geoffrey. *The Riverside Chaucer.* 3rd ed. Ed. Larry D. Benson. Oxford: Oxford UP, 2008.

Chekhov, Anton. "The Lady with the Little Dog." *Stories.* Trans. Richard Pevear and Larissa Volokhonsky. New York: Bantam, 2000. 361–76.

Cicero. *The Nature of the Gods.* Trans. P.G. Walsh. Oxford: Oxford UP, 2008.

———. *The Republic and The Laws.* Trans. Niall Russ. Oxford: Oxford UP, 1998.

Clare, John. "Elegy on the Ruins of Pickworth." *Poems: Description of Rural Life and Scenery.* London, 1820. 65–67.

———. "The Lament of Swordy Well." *Major Works.* Ed. Eric Robinson and David Powell. Oxford: Oxford UP, 1984. 147–52.

Clifton, Lucille. *The Collected Poems of Lucille Clifton 1965–2010.* Ed. Kevin Young and Michael S. Glaser. Rochester: BOA, 2012.

Clute, John, and Peter Nicholls, eds. *The Encyclopedia of Science Fiction.* New York: St. Martin's, 1993.

Cohn, Norman. *Cosmos, Chaos and the World to Come: The Ancient Roots of Apocalyptic Faith.* 2nd ed. New Haven: Yale UP, 2001.

———. *The Pursuit of the Millennium: Revolutionary Millenarians and Mystical Anarchists of the Middle Ages.* Rev. ed. Oxford: Oxford UP, 1970.

Cole, Thomas. *The Course of Empire.* New-York Historical Society. *American Sublime: Landscape Painting in the United States 1820–1880.* By Andrew Wilton and Tim Berringer. Princeton: Princeton UP, 2002. 99, 101, 103, 106, 109.

———. *Thomas Cole's Poetry: The Collected Poems of America's Foremost Painter of the Hudson River School.* Ed. Marshall Tymn. York: George Shumway, 1972.

Coleridge, Samuel Taylor. *The Major Works.* Ed. H.J. Jackson. Oxford: Oxford UP, 1985.

Conniff, Richard. "Lost and Gone Forever." *The New York Times.* 3 Feb. 2011. http://opinionator.blogs.nytimes.com. Accessed 4 Feb. 2011.

Cooper, James Fenimore. *The Leatherstocking Tales, Volume 1.* New York: Library of America, 1985.

Cooper, Susan Fenimore. *Rural Hours.* Syracuse: Syracuse UP, 1968.

Cosslett, Tess. *The "Scientific Movement" and Victorian Literature*. Sussex; Harvester, 1982.

Coupe, Laurence. "Nature/Culture/Gender: Introduction." *The Green Studies Reader: From Romanticism to Ecocriticism*. Ed. Laurence Coupe. London: Routledge, 2000. 119–22.

Crèvecoeur, J. Hector St. John de. *Letters from an American Farmer and Sketches of Eighteenth-Century America*. New York: Penguin, 1981.

Crutzen, Paul J. "Geology of Mankind." *Nature* 415 (Jan. 2002): 23.

Curry, Patrick. *Ecological Ethics: An Introduction*. 2nd ed. Cambridge: Polity, 2011.

Dales, Richard C. *The Scientific Achievement of the Middle Ages*. Philadelphia: U of Pennsylvania P, 1973.

Danielson, Dennis. *The Book of the Cosmos*. Cambridge: Perseus, 2002.

Dante. *The Inferno of Dante*. Trans. Robert Pinsky. New York: Farrar, 1994.

———. *Paradiso*. Trans. Robert Hollander and Jean Hollander. New York: Anchor, 2008.

Darwin, Charles. *The Descent of Man*. 2nd ed. New York: Appleton, 1896.

———. *Notebook C: Transmutation of Species*. Darwin Online. http://darwin-online.org.uk. Accessed 19 July 2015.

———. *On the Origin of Species*. Oxford: Oxford UP, 1996.

Davies, Tony. *Humanism*. 2nd ed. New York: Routledge, 1997.

De Sade, Marquis. *Juliette*. Trans. Austryn Wainhouse. New York: Grove, 1968.

Dean, Bradley P. "Natural History, Romanticism, and Thoreau." In Michael Lewis, 73–90.

DeLoughrey, Elizabeth, and George B. Handley. *Postcolonial Ecologies: Literatures of the Environment*. Oxford: Oxford UP, 2011.

Descartes, René. "Letter to Henry More." *Descartes Selections*. Ed. Ralph M. Eaton. New York: Scribner, 1927. 358–60.

Diamond, Jared. *Collapse: How Societies Choose to Fail or Succeed*. New York: Penguin, 2005.

———. "The Present, Past and Future of Human-Caused Extinctions." *Philosophical Transactions of the Royal Society of London* 325, 1228 (6 Nov. 1989): 469–477.

Dick, Philip K. "Autofac." *The Minority Report and Other Classic Stories*. New York: Citadel, 1987. 1–20.

Dickens, Charles. *The Pickwick Papers*. New York: Everyman's, 1907.

Diderot, Denis. *D'Alembert's Dream*. Trans. Jacques Barzun and Ralph H. Bowen. *Rameau's Nephew and Other Works*. Indianapolis: Hackett, 1956. 89–175.

———. *Diderot on Art Volume II: The Salon of 1767*. Trans. John Goodman. New Haven: Yale UP, 1995.

———. *Thoughts on the Interpretation of Nature and Other Philosophical Works*. Trans. Lorna Sandler. Manchester: Clinamen, 1999.

———. "Articles from the *Encyclopédie.*" *Diderot: Political Writings.* Trans. John Hope Mason and Robert Wolker. Cambridge: Cambridge UP, 1992. 1–30.

Dillard, Annie. "Living Like Weasels." *Teaching a Stone to Talk: Expeditions and Encounters.* New York: Harper, 1982. 11–16.

Disch, Thomas M. *The Genocides.* New York: Vintage, 1965.

Donne, John. *Poetry.* Ed. Donald R. Dickson. New York: Norton, 2007.

Du Bellay, Joachim. *The Regrets: with, The Antiquities of Rome, Three Latin Elegies, and The Defense and Enrichment of the French Language.* Trans. Richard Helgerson. Philadelphia: U of Pennsylvania P, 2006.

Dwight, Timothy. *The Major Poems of Timothy Dwight.* Gainesville: Scholars' Facsimiles, 1969.

Dyer, John. *The Poetical Works of Mark Akenside and John Dyer.* Ed. Robert Aris Willmott. London, 1855.

Edwards, Jonathan. "Sinners in the Hands of an Angry God." *A Jonathan Edwards Reader.* Ed. John E. Smith, Harry S. Stout, and Kenneth P. Minkema. New Haven: Yale UP, 1995. 89–105.

Einstein, Albert. Foreword. *Dialogue Concerning the Two Chief World Systems—Ptolemaic and Copernican.* 2nd ed. Trans. Stillman Drake. Berkeley: U of California P, 1967. vi–xix.

Eiseley, Loren. *The Immense Journey.* New York: Random House, 1957.

Elder, John. "Nature's Refrain in American Poetry." *The Columbia History of American Poetry.* Ed. Jay Parini. New York: Columbia UP, 1993. 707–27.

Emerson, Ralph Waldo. *A Critical Edition of the Major Works.* Ed. Richard Poirier. Oxford: Oxford UP, 1990.

Erdrich, Louise. "Dear John Wayne." *Original Fire: Selected and New Poems.* New York: HarperCollins, 2003. 21–22.

———. "Where I Ought to Be: A Writer's Sense of Place." *New York Times Book Review* (28 July 1985): 1, 23–24.

Evans, Arthur B. "The Origin of SF Criticism: From Kepler to Wells." *SF Studies* 26.2 (1999): 163–86.

Ferber, Linda S. *The Hudson River School: Nature and the American Vision.* New York: Rizzoli, 2009.

Ferber, Michael. *Romanticism: A Very Short Introduction.* Oxford: Oxford UP, 2010.

Fiedler, Leslie A. Introduction. *In Dreams Awake: A Historical-Critical Anthology of SF.* Ed. Leslie A. Fiedler. New York: Laurel, 1975. 11–23.

———. *Love and Death in the American Novel.* New York: Anchor, 1960.

Fisher, Andy. "What is Ecopsychology? A Radical View." *Ecopsychology: Science, Totems, and the Technological Species.* Ed. Peter H. Kahn, Jr. and Patricia H. Hasbach. Cambridge: MIT, 2002. 79–114.

Flannery, Tim. *Here on Earth: A Natural History of the Pla*net. New York: Atlantic Monthly, 2010.

"The Fleshy Part of the Thigh." *The Sopranos.* Dir. Alan Taylor. HBO, 2 April 2006.
Fontenelle, Bernard de. *Conversations on the Plurality of Worlds.* Trans. Elizabeth Gunning. Richmond: Tiger of the Stripe, 2008.
Forster, E.M. "The Machine Stops." *Collected Tales.* New York: Knopf, 1971. 145–97.
Foucault, Michel. *The Order of Things: An Archaeology of Human Sciences.* Trans. Alan Sheridan. New York: Vintage, 1970.
Fowlie, Wallace. *Poem and Symbol: A Brief History of French Symbolism.* University Park: The Pennsylvania SUP, 1990.
Frajlich, Anna. *The Legacy of Ancient Rome in the Russian Silver Age.* Leiden: Brill, 2007.
Freneau, Philip. *Poems.* Ed. Harry Hayden Clark. New York: Hafner, 1929.
Freud, Sigmund. *Beyond the Pleasure Principle.* Trans. A.A. Brill. http://www.bartleby.com. Accessed 16 Aug. 2016.
———. *Reflections on War and Death.* Trans. A.A. Brill. http://www.bartleby.com. Accessed 16 Aug. 2016.
Frost, Robert. *The Poetry of Robert Frost.* Ed. Edward Connery Lathem. New York: Owl, 1969.
Fulford, Tim. "'Nature' Poetry." *The Cambridge Companion to Eighteenth-Century Poetry.* Ed. John Sitter. Cambridge: Cambridge UP, 2001. 109–31.
Furbank, P.N. *Diderot: A Critical Biography.* New York: Knopf, 1992.
Galileo Galilei. *Dialogue Concerning the Two Chief World Systems—Ptolemaic and Copernican.* 2nd ed. Trans. Stillman Drake. Berkeley: U of California P, 1967.
Gallun, Raymond Z. "Seeds of the Dusk." *Catastrophes!* Ed. Isaac Asimov, Martin Harry Greenberg, and Charles G. Waugh. New York: Fawcett, 1981. 226–53.
Garrard, Greg. *Ecocriticism.* 2nd ed. London: Routledge, 2012.
Gay, Peter. *The Enlightenment: The Rise of Modern Paganism.* New York: Norton, 1966.
———. *The Enlightenment: The Science of Freedom.* New York: Norton, 1969.
Gibbon, Edward. *The History of the Decline and Fall of the Roman Empire. Vol.* 3. Ed. Hugh Trevor-Roper. New York: Everyman's, 1994.
Gillispie, Charles Coulston. *The Edge of Objectivity.* Princeton: Princeton UP, 1966.
Gilman, Charlotte Perkins. *Herland and Selected Stories.* Ed. Barbara H. Solomon. New York: Signet, 1992.
Gingerich, Owen. *The Eye of Heaven: Ptolemy, Copernicus, Kepler.* New York: American Institute of Physics, 1993.
Ginsberg, Robert. *The Aesthetics of Ruins.* Amsterdam: Rodopi, 2004.
Glacken, Clarence J. "Man against Nature: An Outmoded Concept." *The Environmental Crisis: Man's Struggle to Live with Himself.* Ed. Harold W. Helfrich, Jr. New Haven: Yale UP, 1970. 127–42.

———. *Traces on the Rhodian Shore: Nature and Culture in Western Thought from Ancient Times to the End of the Eighteenth Century.* Berkeley: U of California P, 1967.

Gleckner, Robert F. *Byron and the Ruins of Paradise.* Baltimore: The Johns Hopkins UP, 1967.

Goethe, Johann Wolfgang von. *Conversations with Eckermann (1823–1832).* Trans. John Oxenford. San Francisco: North Point, 1984.

———. *Faust, Part One.* Trans. David Luke. Oxford: Oxford UP, 1987.

———. *Italian Journey.* Trans. W.H. Auden and Elizabeth Mayer. *Selected Works.* New York: Everyman's, 1999. 383–720.

———. "The Metamorphosis of Plants." *Roman Elegies and Other Poems and Epigrams.* Trans. Michael Hamburger. London: Anvil, 1996. 75–77.

———. "True Enough: To the Physicist." *Roman Elegies and Other Poems and Epigrams.* Trans. Michael Hamburger. London: Anvil, 1996. 116.

Goldstein, Laurence. *Ruins and Empire: The Evolution of a Theme in Augustan and Romantic Literature.* Pittsburgh: U of Pittsburgh P, 1977.

Gordon, Daniel, ed. *Postmodernism and the Enlightenment: New Perspectives in Eighteenth-Century French Intellectual History.* New York: Routledge, 2001.

Gould, Stephen Jay. *The Flamingo's Smile: Reflections in Natural History.* New York: Norton, 1985.

Grainville, Jean-Baptiste Francois Xavier Cousin de. *The Last Man.* Trans. I.F. Clarke and M. Clarke. Middletown: Wesleyan UP, 2002.

Gray, John. *Straw Dogs: Thoughts on Humans and Other Animals.* New York: Farrar, 2002.

Grayling, A.C. *Ideas That Matter: The Concepts That Shape the 21st Century.* New York: Basic, 2010.

Greenblatt, Stephen. *The Swerve: How the World Became Modern.* New York: Norton, 2011.

Groom, Nick. *The Gothic: A Very Short Introduction.* Oxford: Oxford UP, 2012.

Guthrie, W.K.C. *The Greek Philosophers: From Thales to Aristotle.* New York: Harper, 1950.

Gutting, Gary. "On Experts and Global Warming." *The New York Times* 12 July 2011. http://opinionator.blogs.nytimes.com. Accessed 12 July 2011.

Guyer, Paul. *Kant.* London: Routledge, 2006.

Hadot, Pierre. *What Is Ancient Philosophy?* Trans. Michael Chase. Cambridge: Belknap, 2004.

Hage, Erik. *Cormac McCarthy: A Literary Companion.* Jefferson: McFarland, 2010.

Hallam, Tony. *Catastrophes and Lesser Calamities: The Causes of Mass Extinctions.* Oxford: Oxford UP, 2005.

Hamilton, Edmond. "The Earth-Owners." *Weird Tales* (Aug. 1931): 22–31, 138–40.

Hankins, Thomas. *Science and the Enlightenment.* Cambridge: Cambridge UP, 1985.

Hardin, Garrett. "The Tragedy of the Commons." *Science* 162 (1968): 1243–48.

Hardy, Thomas. "The Convergence of the Twain." *Selected Poetry.* Oxford: Oxford UP, 1996. 72–73.

Harjo, Joy. *How We Became Human: New and Selected Poems: 1975–21.* New York: Norton, 2002.

Harrison, Robert Pogue. *Forests: The Shadow of Civilization.* Chicago: U of Chicago P, 1992.

Harvey, Mark. "Loving the Wilderness in Postwar America." In Michael Lewis, 187–203.

Hawthorne, Nathaniel. *The Marble Faun.* Oxford: Oxford UP, 2002.

———. *Tales.* Ed. James McIntosh. New York: Norton, 1987.

Hedges, Chris. *Empire of Illusion: The End of Literacy and the Triumph of Spectacle.* New York: Nation, 2009.

Heidegger, Martin. "Letter on Humanism." Trans. Frank A. Capuzzi and J. Glenn Gray. *Basic Writings.* Ed. David Farrell Krell. New York: Harper, 2008. 213–67.

Herbert, George. *Works.* Ed. F.E. Hutchinson. Oxford: Clarendon, 1941.

Hitchens, Christopher. *Arguably: Essays by Christopher Hitchens.* New York: Hachette, 2011.

Hodgson, William Hope. *The House on the Borderland.* Mineola: Dover, 2008.

Hogan, Linda. *Dwellings: A Spiritual History of the Living World.* New York: Touchstone, 1995.

Holmes, Richard. *The Age of Wonder: The Romantic Generation and the Discovery of the Beauty and Terror of Science.* New York: Vintage, 2010.

———. *The Pursuit.* Glasgow: Flamingo, 1974.

Hood, Thomas. "The Last Man." *Selected Poems.* Ed. John Clubbe. Cambridge: Harvard UP, 1970. 134–41.

Hughes, Robert. *American Visions: The Epic History of Art in America.* New York: Knopf, 1997.

Hulme, T.E. "Modern Art and Its Philosophy." *Speculations.* Ed. Herbert Read. New York: Harcourt, 1924. 73–109.

"Humans nearly extinct 70,000 years ago." *The Toronto Star.* 24 Apr. 2008. https://www.thestar.com. Accessed. 20 Dec. 2015.

Hume, David. *Dialogues Concerning Natural Religion.* Ed. Martin Bell. London: Penguin, 1990.

Hutton, James. *Theory of the Earth.* http://pages.uwc.edu. Accessed 20 Aug. 2016.

Huxley, Aldous. *Brave New World.* New York: Harper, 1932.

———. *Literature and Science.* New York: Harper, 1963.

Introduction to Global Change. http://www.globalchange.umich.edu. Accessed 8 July 2016.

Inwood, Brad, and L.P. Gerson, trans. *The Stoics Reader: Selected Writings and Testimonia.* Indianapolis: Hackett, 2008.
Irving, Washington. *A History of New York.* New York: Penguin, 2008.
———. "The Legend of Sleepy Hollow." *History, Tales and Sketches.* New York: Library of America, 1983. 1058–88.
Israel, Jonathan. *A Revolution of the Mind: Radical Enlightenment and the Intellectual Origins of Modern Democracy.* Princeton: Princeton UP, 2011.
James, Henry. *Daisy Miller: A Study. Tales of Henry James.* 2nd ed. New York: Norton, 2003. 3–51.
———. *Italian Hours. Collected Travel Writings: The Continent.* New York: Library of America. 279–619.
Jardine, Lisa. *Ingenious Pursuits: Building the Scientific Revolution.* New York: Nan A. Talese, 1999.
Jefferies, Richard. *After London, or Wild England.* Charleston: BiblioBazaar, 2006.
Jeffers, Robinson. *Collected Poetry.* Ed. Tim Hunt. Stanford: Stanford UP, 1988–1991. 3 vols.
———. Preface. *The Double Axe and other Poems.* New York: Random, 1948. vii–viii.
Johnson, Benjamin. "Wilderness Parks and Their Discontents." In Michael Lewis, 113–30.
Joshi, S.T. *H.P. Lovecraft: A Life.* West Warwick: Necronomicon, 1996.
Kahn, Charles. "The Achievement of Early Greek Philosophy: A Drama in Five Acts from Thales to the Timaeus." *Early Greek Philosophy: The Presocratics and the Emergence of Reason.* Ed. Joe McCoy. Catholic U of America P, 2013. 1–17.
Kant, Immanuel. *The Critique of Judgement.* Trans. James Creed Meredith. Oxford: Clarendon, 1952.
———. "What Is Enlightenment?" Trans. Peter Gay. *Introduction to Contemporary Civilization in the West.* New York: Columbia UP, 1954. 1071–76.
Karman, James. *Robinson Jeffers: Poet of California.* Rev. ed. Brownsville: Story Line, 1995.
Keller, Evelyn Fox. "Secrets of God, Nature, and Life." *The Gender and Science Reader.* Ed. Muriel Lederman and Ingrid Bartsch. London: Routledge, 2001. 98–110.
Kinsley, William. "Apocalypse." *A Dictionary of Biblical Tradition in English Literature.* Ed. David L. Jeffrey. Grand Rapids: Eerdmans 1992. 47–48.
Kolodny, Annette. *The Lay of the Land: Metaphor as Experience and History in American Life and Letters.* Chapel Hill: The U of North Carolina P, 1975.
Kroeber, Karl. *Romantic Fantasy and Science Fiction.* New Haven: Yale UP, 1988.
Kumin, Maxine. "Intimations of Mortality." *Women's Review of Books* 10 (7 Apr. 1993): 19.

Kunsa, Ashley. "'Maps of the World in Its Becoming': Post-Apocalyptic Naming in Cormac McCarthy's *The Road*." *Journal of Modern Literature* 33.1 (2009): 57–74.

L'Avventura. Dir. Michelangelo Antonioni. DVD. Cino del Duca, 1960.

Laforgue, Jules. *Selected Writings*. Trans. William Jay Smith. New York: Grove, 1956.

Lambert, Zuidervaart. "Adorno." 26 Oct. 2015. *Stanford Encyclopedia of Philosophy*. https://plato.stanford.edu. Accessed 15 May 2016.

Larson, Sidner. *Captured in the Middle: Tradition and Experience in Contemporary Native American Writing*. Seattle: U of Washington P, 2000.

The Last Wave. Dir. Peter Weir. DVD. Australian Film Commission, 1977.

Latour, Bruno. *We Have Never Been Modern*. Trans. Catherine Porter. Cambridge: Harvard UP, 1993.

Le Guin, Ursula K. "Direction of the Road." *The Wind's Twelve Quarters*. New York: Perennial, 1976. 267–74.

———. Introduction. H.G. Wells. *The Time Machine: An Invention*. New York: Modern Library, 2002. xi–xvi.

———. "The New Atlantis." *Catastrophes!* Ed. Isaac Asimov, Martin Harry Greenberg, and Charles G. Waugh. New York: Fawcett, 1981. 194–217.

———. *The Word for World Is Forest*. New York: Tor, 2010.

Leinster, Murray. *Mad Planet. Under the Moons of Mars: A History and Anthology of "The Scientific Romance" in the Munsey Magazines, 1912–1920*. Ed. Sam Moskowitz. New York: Holt, 1970. 221–68.

Lem, Stanislaw. *The Futurological Congress*. Trans. Michael Kandel. San Diego: Harvest, 1974.

———. "Let Us Save the Universe (An Open Letter from Ijon Tichy)." *Memoirs of a Space Traveler: Further Reminiscences of Ijon Tichy*. Trans. Joel Stern and Maria Swiecicka-Ziemianek. San Diego: Harvest, 1982. 139–53.

Leopardi, Giacomo. "La Ginestra." *Conti*. Trans. Jonathan Galassi. London: Penguin, 2010. 286–309.

Leopold, Aldo. *A Sand County Almanac: With Essays on Conservation from Round River*. New York: Ballantine, 1970.

Lewis, C.S. *The Discarded Image: An Introduction to Medieval and Renaissance Literature*. Cambridge: Canto, 1964.

———. "On Science Fiction." *Of Other Worlds: Essays and Stories*. Ed. Walter Hooper. San Diego: Harvest, 1966. 59–73.

Lewis, Michael, ed. *American Wilderness: A New History*. Oxford: Oxford UP, 2007.

———. "Wilderness and Conservation Science." In Michael Lewis, 205–22.

Life After People. Dir. David de Vries. DVD. A&E Home Video, 2008.

Ligotti, Thomas. *The Conspiracy against the Human Race: A Contrivance of Horror*. New York: Hippocampus, 2010.

Lincoln, Kenneth. *Cormac McCarthy: American Canticles*. New York: Palgrave Macmillan, 2009.
Little, Charles E. "Books for the Wilderness." *Wilderness* (Summer 1988): 56, 59.
"Living Planet Index." 2016. WWF Global. http://wwf.panda.org. Accessed 13 Dec. 2016.
London, Jack. *The Scarlet Plague*. London: Hesperus, 2008.
"The Long Bright Dark." *True Detective*. Dir. Cary Joji Fukunaga. HBO, 12 Jan. 2014.
Lonsdale, Roger. "Introduction." *Gray and Collins: Poetical Works*. Ed. Roger Lonsdale. Oxford: Oxford UP, 1977. xi–xxiii.
Lovecraft, H.P. "The Colour out of Space." *The Annotated Lovecraft*. Ed. S.T. Joshi. New York: Dell, 1997. 57–100.
Luchte, James. "Heidegger's 'Letter on Humanism'-A Reading." https://luchte.wordpress.com. Accessed 5 July 2016.
Lucian. *Selected Dialogues*. Trans. C.D.N. Costa. Oxford: Oxford UP, 2009.
Lucretius. *On the Nature of Things: De rerum natura*. Trans. Anthony M. Esolen. Baltimore: Johns Hopkins UP, 1995.
Macrobius. *Commentary on the Dream of Scipio*. Trans. William Harris Stahl. New York: Columbia UP, 1952.
Makarius, Michel. *Ruins*. Paris: Flammarion, 2005.
Marchand, Leslie A. *Byron: A Portrait*. New York: Knopf, 1970.
Marsh, George Perkins. *So Great a Vision: The Conservation Writings of George Perkins Marsh*. Hanover: Middlebury College P, 2001.
Mather, Cotton. *The Christian Philosopher*. Ed. Winston U. Solberg. Urbana: U of Illinois P, 1994.
Maupassant, Guy de. "The Horla." *Selected Short Stories*. Trans. Roger Colet. London: Penguin, 1971. 313–44.
McCann, Jerome J. *Fiery Dust: Byron's Poetic Development*. Chicago: U of Chicago P, 1968.
McCarthy, Cormac. *Blood Meridian: Or, The Evening Redness in the West*. New York: Vintage, 1985.
———. *Child of God*. New York: Vintage, 1973.
———. Interview with Ira Flatow. *Science Friday*. National Public Radio. 8 April 2011.
———. *No Country for Old Men*. New York: Vintage, 2005.
———. *Outer Dark*. New York: Vintage, 1968.
———. *The Road*. New York: Vintage, 2006.
———. *The Sunset Limited*. New York: Vintage, 2006.
McGuire, Bill. *A Guide to the End of the World: Everything You Never Wanted to Know*. Oxford: Oxford UP, 2002.

McLean, W. Scott, Eldridge M. Moores, and David A. Robinson. "Poetry Matters: Gary Snyder." *Language of the Earth.* 2nd ed. Ed. Frank H.T. Rhodes, Richard O. Stone, and Bruce D. Malamud. Malden: Blackwell, 2009. 238–41.

McPherson, Guy. "Resources and Anthropocentrism." https://guymcpherson.com. 12 Oct. 2009. Accessed 19 Dec. 2015.

Melancholia. Dir. Lars von Triers. DVD. Zentropa, 2011.

Melville, Herman. "Bartleby, the Scrivener." *The Piazza Tales and Other Prose Pieces 1839–1860.* Ed. Harrison Hayford, et al. Evanston: Northwestern UP, 1987. 13–45.

———. *Moby-Dick.* 2nd ed. Ed. Hershel Parker and Harrison Hayford. New York: Norton, 2002.

Merchant, Carolyn. *The Death of Nature: Women, Ecology and the Scientific Revolution.* New York: Harper, 1980.

Meredith, George. "Meditation under Stars." *Works.* Vol. XXV. London: Constable, 1910. 265–67.

Midgely, Mary. "The End of Anthropocentrism?" *Philosophy and the Natural Environment.* Ed. Robin Attfield and Andrew Belsey. Cambridge: Cambridge UP, 1994. 103–12.

Miller, Angela. "The Fate of Wilderness in American Landscape Art: The Dilemmas of 'Nature's Nation.'" In Michael Lewis, 91–112.

Miller, J. Hillis. "Friedrich Schlegel: Catachreses for Chaos." *Others.* Princeton: Princeton UP, 2001. 5–42.

Molière. *The Misanthrope and Tartuffe.* Trans. Richard Wilbur. New York: Mariner, 1965.

Moltmann, Jürgen. "Christianity—Political Theology." In Bron Taylor, 355–357.

———. *God in Creation: A New Theology of Creation and the Spirit of God.* Trans. Margaret Kohl. Minneapolis: Fortress, 1993.

Momaday, N. Scott. "An Interview by Joseph Bruchac." *The American Poetry Review* 13.4 (1984): 13–18.

Moore, Bryan L. *Ecology and Literature: Ecocentric Personification from Antiquity to the Twenty-first Century.* New York: Palgrave Macmillan, 2008.

Moore, Marianne. *Complete Poems.* New York: Viking, 1981.

Naitum, Duane, ed. *Harper's Anthology of 20th Century Native American Poetry.* New York: Harper, 1987.

Nash, Roderick. *Wilderness and the American Mind.* 4th ed. New Haven; Yale UP, 2001.

Nashe, Thomas. "Litany in Time of Plague." *The Oxford Anthology of English Literature Vol. 1.* Ed. Frank Kermode and John Hollander. New York: Oxford UP, 1973. 614–15.

Nerval, Gerard de. "El Desdichado." Trans. Richard Lattimore. *The Anchor Anthology of French Poetry: From Nerval to Valéry in English Translation.* Ed. Angel Flores. New York: Anchor, 1958. 9.

———. "Sylvie." *Selected Writings.* Trans. Richard Sieburth. London: Penguin, 1999. 145–77.

"New Genome Comparison Finds Chimps, Humans Very Similar at the DNA Level." NIH News. 31 Aug. 2005. https://www.genome.gov. Accessed 12 July 2016.

Nichols, Ashton. Introduction. *Romantic Natural Histories: William Wordsworth, Charles, Darwin, and Others.* Ed. Ashton Nichols. Boston: Houghton, 2004. 1–20.

Nietzsche, Friedrich. *The Antichrist.* Trans. Walter Kaufmann. *The Portable Nietzsche.* New York: Penguin, 1954. 565–656.

———. *Beyond Good and Evil.* Trans. R.J. Hollingdale. London: Penguin, 1990.

———. *Human, All Too Human: A Book for Free Spirits.* 2nd ed. Trans. R. J. Hollingdale. Cambridge: Cambridge UP, 1996.

———. *On the Genealogy of Morals.* Trans. Walter Kaufmann. *Basic Writings of Nietzsche.* New York: Modern Library, 2000. 437–599.

Oelschlaeger, Max. *The Idea of Wilderness: From Prehistory to the Age of Ecology.* New Haven: Yale UP, 1991.

Oliver, Mary. *New and Selected Poems Volume One.* Boston: Beacon, 1992.

Oreskes, Naomi, and Erik M. Conway*The Collapse of Western Civilization: A View from the Future.* New York: Columbia UP, 2014.

Orwell, George. *1984.* New York: Plume, 1983.

Osler, Margaret J. *Reconfiguring the World: Nature, God, and Human Understanding from the Middle Ages to Early Modern Europe.* Baltimore: Johns Hopkins UP, 2010.

Paine, Thomas. *The Age of Reason.* Ed. Moncure Daniel Conway. N.p: Merchant, 2010.

Paley, Morton D. *Apocalypse and Millennium in English Romantic Poetry.* Rev. ed. Oxford: Oxford UP, 1999.

Palmer, John "Parmenides." 3 June 2012. Stanford Encyclopedia of Philosophy. https://plato.stanford.edu. Accessed 5 July 2015.

"Parmenides." *The Concise Oxford Companion to Classical Literature.* Ed. M.C. Howatson and Ian Chilvers. Oxford: Oxford UP, 1993. 396–97.

Peterson, Anna L. *Being Human: Ethics, Environment, and Our Place in the World.* Berkeley: U of California P, 2001.

Petrarch, Francesco. *Letters on Familiar Matters (Rerum Familiarium Libri): Vol. 1: Books I–VIII.* Trans. Aldo S. Bernardo. New York: Italica, 2005.

Philipse, Herman. *Heidegger's Philosophy of Being.* Princeton: Princeton UP, 1998.

Pimm, S.L., et al. "The biodiversity of species and their rates of extinction, distribution, and protection." *Science* 30 May 2014. http://science.sciencemag.org. Accessed 15 May 2015.

Pite, Ralph. "Wordsworth and the Natural World." *The Cambridge Companion to Wordsworth.* Ed. Stephen Gill. Cambridge: Cambridge UP, 2003. 180–195.

Plato. *Complete Works*. Ed. John M. Cooper. Indianapolis: Hackett, 1997.

Pliny the Elder. *Natural History: A Selection*. Trans. John F. Healy. London: Penguin, 1991.

Plumwood, Val. "Androcentrism and Anthrocentrism: Parallels and Politics." *Ethics and the Environment* 1.2 (1996): 119–52.

———. *Environmental Culture: The Ecological Crisis of Reason*. London: Routledge, 2002.

———. *Feminism and the Mastery of Nature*. London: Routledge. 1993.

Pocock, J.G.A. "Gibbon's*Decline and Fall* and the World View of the Late Enlightenment." *Virtue, Commerce, and History: Essays on Political Thought and History, Chiefly in the Eighteenth Century*. Ed. J.G.A. Pocock. Cambridge: Cambridge UP, 1985. 143–56.

Poe, Edgar Allan. "Al Aaraaf." *Complete Poems*. Ed. Thomas Olive Mabbott. Urbana: U of Illinois P, 2000. 92–127.

———. "The Colloquy of Monos and Una." *Tales and Sketches Vol. 1: 1831–1842*. Ed. Thomas Ollive Mabbott. Urbana: U of Illinois P, 2000. 607–19.

———. "The Conversation of Eiros and Charmion." *Tales and Sketches Vol. 1: 1831–1842*. Ed. Thomas Ollive Mabbott. Urbana: U of Illinois P, 2000. 451–62.

Pohl, Frederik. "Creation Myths of the Recently Extinct." *Platinum Pohl: The Collected Best Stories*. New York: Tor, 2005. 407–08.

———. Headnote. "The Remorseful." *The Best of C.M. Kornbluth*. Ed. Frederik Pohl. New York: Ballantine, 1976. 100.

———, and Cyril Kornbluth. *The Space Merchants*. New York: St. Martin's, 1952.

Polybius. *The Histories of Polybius. Vol. 6*. Trans. W. R. Paton. Cambridge: Harvard UP, 1927.

Pope, Alexander. *The Major Works*. Ed. Pat Rogers. Oxford: Oxford UP, 2006.

Pound, Ezra. "E.P. Ode pour l'Election de Son Sepulchre." *Hugh Selwyn Mauberley. New Selected Poems and Translations*. Ed. Richard Sieburth. New York: New Directions, 2010. 111–14.

Quinn, Daniel. *Ishmael: An Adventure of the Mind and Spirit*. New York: Bantam, 1992.

Quintilian. *Institutio Oratoria*. Trans. H.E. Butler. 4 vols. Loeb Classical Library. Cambridge UP, 1920–22.

The Qu'ran. Sahih International. 2009-11. http://corpus.quran.com/translation. Accessed 12 Aug. 2016.

Red Jacket. "Reply to the Missionary Jacob Cram." *The Norton Anthology: American Literature Shorter 8th ed. Vol. 1*. Ed. Nina Baym. New York: Norton, 2013. 230–31.

Rees, Sir Martin. "Templeton Prize 2011: Full transcript of Martin Rees's acceptance speech." *The Guardian* 6 Apr. 2011. https://www.theguardian.com. Accessed 17 Nov. 2016.

Rexroth, Kenneth. Author's Note to "Thou Shalt Not Kill." *The Portable Beat Reader*. Ed. Ann Charters. New York: Penguin, 1992. 233.
———. *Complete Poems*. Ed. Sam Hamill and Bradford Morrow. Port Townsend: Copper Canyon, 23.
———. "Poets, Old and New." *Assays*. New York: New Directions, 1961. 206–39.
Rhodes, Frank H.T., Richard O. Stone, and Bruce D. Malamud, eds. "Poetry Matters: Gary Snyder." *Language of the Earth: A Literary Anthology*. Malden: Blackwell, 2008. 238–41.
Robinson, Eric, David Powell, and Margaret Grainger. Introduction. *The Early Poems of John Clare, 1804–1822: Vol. 1*. Oxford: Oxford UP, 1989. ix–xxiv.
Robinson, Kim Stanley. *Forty Signs of Rain*. New York: Bantam, 2004.
Rogers, Pattiann. "Twentieth Century Cosmology and the Soul's Habitation." *The Measured Word: On Poetry and Science*. Ed. Kurt Brown. Athens: U of Georgia P, 2001. 1–13.
Rousseau, Jean-Jacques. *Reveries of the Solitary Walker*. Trans. Russell Goulbourne. Oxford: Oxford UP, 2011.
Rowlandson, Mary. *A Narrative of the Captivity and Restoration of Mrs. Mary Rowlandson*. *The Norton Anthology: American Literature Vol. A*. 8th ed. Ed. Nina Baym. New York: Norton, 2012. 57–88.
Russ, Joanna. *The Female Man*. Boston: Beacon, 1975.
Russell, Bertrand. "A Free Man's Worship." *Mysticism and Logic*. Mineloa: Dover, 2004: 36–44.
Sartre, Jean Paul. *Existentialism Is a Humanism*. Trans. Carol Macomber. New Haven: Yale UP, 2007.
Schuyler, David. *Sanctified Landscape: Writers, Artists, and the Hudson River Valley, 1820–1909*. Ithaca: Cornell UP, 2012.
Schuyler, Robert Livingston. "Man's Greatest Illusion." *Proceedings of the American Philosophical Society* 92.1 (1948): 46–50.
Seneca. *Natural Questions*. Trans. Harry M. Hine. Chicago: U of Chicago P, 2010.
Shaftesbury, Anthony, Earl of. *Characteristics of Men, Manners, Opinions, Times*. Ed. John M. Robertson. Indianapolis: Bobbs-Merrill, 1964.
Shakespeare, William. *Complete Works*. 2nd ed. Ed. John Jowett, et al. Oxford: Oxford UP, 1986.
Shapin, Steven. *The Scientific Revolution*. Chicago: U of Chicago P, 1998.
Shelley, Mary. *Frankenstein 1818 Text*. Oxford: Oxford UP, 1994.
———. *The Last Man*. Oxford: Oxford UP, 1994.
Shelley, Percy Bysshe. *Shelley's Poetry and Prose*. 2nd ed. Ed. Donald H. Reiman and Neil Fraistat. New York: Norton, 2002.
———. *Shelley's Prose; or, The Trumpet of a Prophecy*. Ed. David Lee Clark. Albuquerque: U of New Mexico P, 1954.
Shiel, Matthew P. *The Purple Cloud*. Boston: G.K. Hall, 1977.

Sigourney, Lydia Huntley. "Fallen Forests." *Selected Poetry and Prose*. Ed. Gary Kelly. Peterborough: Broadview, 2008. 243–44.

Silko, Leslie Marmon. *Ceremony*. New York: Penguin, 1977.

Simmel, Georg. "The Ruin." Trans. David Kettler. *Georg Simmel, 1858–1918: A Collection of Essays*. Ed. Kurt H. Wolff. Columbus: Ohio SUP, 1959. 259–66.

Slonczewski, Joan, and Michael Levy. "Science Fiction and the Life Sciences." *The Cambridge Companion to Science Fiction*. Ed. Edward James and Farah Mendlesohn. Cambridge: Cambridge UP, 2003. 174–85.

Smith, Charlotte. *Beachy Head*. *The Norton Anthology of English Literature: The Romantic Period Vol. D*. Ed. Stephen Greenblatt. New York: Norton, 2012. 59–77.

Snyder, Gary. "A Letter to Ed Abbey." *Resist Much, Obey Little: Remembering Ed Abbey*. Ed. James R. Hepworth and Gregory McNamee. San Francisco: Sierra, 1996. 182–88.

———. *Mountains and Rivers Without End*. Washington, D.C.: Counterpoint, 1996.

———. *No Nature: New and Selected Poems*. New York: Pantheon, 1992.

———. *Turtle Island*. New York: New Directions, 1974.

Somerville, Mary. *On the Connexion of the Physical Sciences*. https://archive.org/. Accessed 9 Dec. 2016.

Soper, Kate. *Humanism and Anti-Humanism*. London: Open Court, 1986.

Sophocles. *Antigone, Oedipus the King and Electra*. Trans. H.D.F. Kitto. Oxford: Oxford UP, 1994.

"Speed of glacier retreat worldwide 'historically unprecedented', says report." *The Guardian*. 4 Aug. 2015. www.theguardian.com. Accessed 7 Dec. 2016.

Spenser, Edmund. *Complaints*. *The Yale Edition of the Shorter Poems of Edmund Spenser*. Ed. Willam A. Oram, et al. New Haven: Yale UP, 1989. 215–457.

———. *The Faerie Queene*. Ed. A.C. Hamilton. London: Longman, 1977.

Speth, James Gustave. *The Bridge at the Edge of the World: Capitalism, the Environment, and Crossing from Crisis to Sustainability*. New Haven: Yale UP, 2008.

Spinoza, Baruch. *On the Improvement of the Understanding. The Ethics. Correspondence*. Trans. R.H.M. Elwes. New York: Dover, 1955.

Stableford, Brian. "Ecology." *The Encyclopedia of Science Fiction*. Ed. John Clute and Peter Nicholls. New York: St. Martin's, 1993. 365–66.

Staël, Madame de. *Corrine, or Italy*. Trans. Sylvia Raphael. Oxford: Oxford UP, 1998.

Starobinski, Jean. *Blessings in Disguise, Or, The Morality of Evil*. Cambridge: Harvard UP, 1993.

Steiner, Gary. *Anthropocentrism and Its Discontents: The Moral Status of Animals in the History of Western Philosophy*. Pittsburgh: U of Pittsburgh P, 2005.

Stewart, George R. *Earth Abides*. New York: Del Rey, 2006.

Straub, Peter. Foreword. H.G. Wells. *The Island of Dr. Moreau*. New York: Modern Library, 2002. xi–xxi.

Stoll, Steven. "Farm against Forest." In Michael Lewis, 55–72.

Sutter, Paul. "Putting Wilderness in Context: The Interwar Origins of the Modern Wilderness Idea." In Michael Lewis, 167–86.

Suvin, Darko. "Estrangement and Cognition." 24 Nov. 2014. Strangehorizons.com. Accessed 19 Aug. 2016.

Taylor, Alan. *American Colonies: The Settling of North America*. New York: Penguin, 2001.

Taylor, Bron, ed. *The Encyclopedia of Religion and Nature*. 2 vols. New York: Continuum, 2005, 2008.

Taylor, Paul W. *Respect for Nature: A Theory of Environmental Ethics*. Princeton: Princeton UP, 1986.

Tennyson, Alfred. *Tennyson's Poetry*. 2nd ed. Ed. Robert W. Hill, Jr. New York: Norton, 1999.

"A tenth of the world's wilderness lost since the 1990s." *ScienceDaily*. 8 Sep. 2016. www.sciencedaily.com. Accessed 8 Sep. 2016.

Thomson, James. "The City of Dreadful Night." *Poems and Some Letters of James Thomson*. Ed. Anne Ridler. Carbondale: Southern Illinois UP, 1963. 177–205.

———. "Proposals for the Extinction of Evil and Misery." *The Speedy Extinction of Evil and Misery: Selected Prose of James Thomson (B.V.)*. Ed. William David Schaefer. Berkeley: U of California P, 1967. 15–33.

Thoreau, Henry David. *Correspondences*. Ed. Walter Harding and Carl Bode. New York: New York UP, 1958.

———. *Journal, Volume 5*. Ed. Patrick F. O'Connell Princeton: Princeton UP, 1997.

———. *Walden*. Princeton: Princeton UP, 1971.

———. "Walking." *Excursions*. Ed. Joseph J. Moldenhauer. Princeton: Princeton UP, 2007. 185–222.

Tiptree, James, Jr. "The Girl Who Was Plugged In." *Her Smoke Rose Up Forever*. San Francisco: Tachyon, 2004. 43–78.

Trilling, Lionel. "The Greatness of *Huckleberry Finn*." Mark Twain. *Adventures of Huckleberry Finn*. Ed. Sculley Bradley, Richard Croom Beatty, and E. Hudson Long. New York: Norton, 1961. 310–20.

Tucker, Mary Evelyn. "Confucianism." In Bron Taylor, 407–10.

Turner, James Morton. "The Politics of Modern Wilderness." In Michael Lewis, 243–62.

Twain, Mark. *Adventures of Huckleberry Finn*. Ed. Sculley Bradley, Richard Croom Beatty, and E. Hudson Long. New York: Norton, 1961.

———. *Autobiography of Mark Twain: The Complete and Authoritative Edition, Vol. 1*. Ed. Harriet Elinor Smith. Berkeley: U of California P, 2010.

———. *Letters from the Earth: Uncensored Writings.* Ed. Bernard DeVoto. New York: Harper, 1962.

Valéry, Paul. "The Crisis of the Mind." *An Anthology.* Trans. Denise Folliot and Jackson Matthews. Princeton: Princeton UP, 1977. 94–107.

Verlaine, Paul. "Apathy" ("Langueur"). Trans. Muriel Kittel. *The Anchor Anthology of French Poetry: From Nerval to Valéry in English Translation.* Ed. Angel Flores. New York: Anchor, 1958. 100–01.

Verne, Jules. "The Eternal Adam." Trans. I.O. Evans. *The Eternal Adam and Other Stories.* London: Phoenix, 1999. 211–44.

Volney, Constantin Francois de. *The Ruins or Meditation on the Revolutions of Empires and the Law of Nature.* Middlesex: Echo, 2007.

Voltaire. *Candide, Zadig and Selected Stories.* Trans. Donald Frame. New York: Signet, 1961.

———. "Dialogues between Lucretius and Posidonius, First Colloquy." *The Works of Voltaire: Romances Vol. III.* Trans. William F. Fleming. Paris, 1901. 76–85.

———. *The Lisbon Earthquake.* Trans. Tobias Smollett. *The Portable Voltaire.* New York: Penguin, 1949. 556–69.

———. *Philosophical Letters.* Trans. Ernest Dilworth. Indianapolis: Bobbs-Merrill, 1961.

The Voluntary Human Extinction Movement. http://www.vhemt.org/. Accessed 12 Dec. 2015.

Vonnegut, Kurt, Jr. "The Big Space Fuck." *Palm Sunday.* New York: Dial, 2011. 207–14.

———. *The Sirens of Titan.* New York: Dial, 1998.

Wakoski, Diane. "Joining the Visionary 'Inhumanists'." *Resist Much, Obey Little: Remembering Ed Abbey.* Ed. James R. Hepworth and Gregory McNamee. San Francisco: Sierra, 1996. 168–74.

The Walking Dead. Created by Frank Darabont. AMC, 2010-present.

WALL-E. Dir. Andrew Stanton. DVD. Disney, 2008.

Wallace, Alfred Russel. *Darwinism.* https://en.wikisource.org/wiki/Darwinism_(Wallace). Accessed 22 Aug. 2016.

Wallace, David Rains. "What Ever Happened to Robinson Jeffers?" Rev. of*The Collected Poetry of Robinson Jeffers.* Ed. Tim Hunt. *Los Angeles Times* 29 Oct 2000. http://articles.latimes.com/2000/oct/29/books/bk-43643. Accessed 26 July 2015.

Walls, Laura Dassow. *Emerson's Life in Science: The Culture of Truth.* Ithaca: Cornell UP, 2003.

Waterfield, Robin, trans. *The First Philosophers.* Oxford: Oxford UP, 2000.

Watkins, John. "Spenser's Poetry and the Apocalypse." *Early Modern English Poetry: A Critical Companion.* Oxford: Oxford UP, 2007. 90–101.

Weisman, Alan. *The World without Us.* New York: Picador, 2007.

Wells, H.G. "A Dream of Armageddon." *Complete Stories.* Ed. John Hammond. London: Phoenix, 1998. 547–67.

———. "The Empire of the Ants." *Complete Stories.* Ed. John Hammond. London: Phoenix, 1998. 585–97

———. *The Island of Dr. Moreau.* New York: Modern Library, 2002.

———. "The Man of the Year Million." *Journalism and Prophecy 1893–1946.* Ed. W. Warren Wagar. Boston: Houghton, 1964. 3–9.

———. *Mind at the End of Its Tether and The Happy Turning.* New York: Didier, 1946.

———. *The Outline of History: Being a Plain History of Life and Mankind.* Garden City: Garden City, 1920.

———. "The Star." *Complete Stories.* Ed. John Hammond. London: Phoenix, 1998. 281–89.

———. *The Time Machine: An Invention.* New York: Modern Library, 2002.

———. *The War of the* Worlds. London, Penguin, 2005.

Wheatley, Phillis. "Thoughts on the Works of Providence." *Complete Writings.* Ed. Vincent Carretta. New York: Penguin, 2001. 26–29.

White, E.B. "The Morning of the Day They Did It." *Past, Present, and Future Perfect: A Text Anthology of Speculative and Science Fiction.* Ed. Gregory Fitz Gerald and Jack C. Wolf. Greenwich: Fawcett, 1973. 131–50.

———. "Sootfall and Fallout." *Essays of E.B. White.* New York: Harper, 1977. 112–24.

Whitman, Walt. *Song of Myself. Leaves of Grass and Other Writings.* Ed. Michael Moon. New York: Norton, 2002. 26–78.

Wigglesworth, Michael. *The Poems of Michael Wigglesworth.* Ed. Ronald A. Bosco. Lanham: UP of America, 1989.

Wild, Peter. *Pioneer Conservationists of Western America.* Missoula: Mountain Press, 1979.

Wilhelm, Kate. *Where Late the Sweet Birds Sang.* New York: Orb, 1976.

Williams, Raymond. *The Country and the City.* New York: Oxford UP, 1973.

Wilson, E.O. *The Creation: An Appeal to Save Life on Earth.* New York: Norton, 2006.

Wilson, Edmund. *Axel's Castle: A Study of the Imaginative Literature of 1870–1930.* New York: Farrar, 1931.

Wilton, Andrew, and Tim Berringer. *American Sublime: Landscape Painting in the United States 1820–1880.* Princeton: Princeton UP, 2002.

Wineapple, Brenda. *Hawthorne: A Life.* New York: Knopf, 2003.

Winthrop, John. *A Model of Christian Charity. The Norton Anthology: American Literature Vol. A.* 8th ed. Ed. Nina Baym. New York: Norton, 2012. 166–77.

———. *Winthrop Papers Volume 4: 1648–1644.* Boston: Merrymount, 1944.

Wordsworth, William. *Selected Poems and Prefaces.* Ed. Jack Stillinger. Boston: Riverside, 1965.

———, and Samuel Taylor Coleridge. *Lyrical Ballads.* 2nd ed. Ed. Michael Mason. Harlow: Pearson, 2007.
Worster, Donald. "Epilogue: Nature, Liberty, and Equality." In Michael Lewis, 263–72.
Wulf, Andrea. *The Invention of Nature: Alexander von Humboldt's New World.* New York: Vintage, 2015.
Zamyatin, Yevgeny. *We.* Trans. Natasha Randall. New York: Modern Library, 2006.
Zeller, Tom. "Scientists Agree (Again): Climate Change Is Happening." The Huffington Post. 16 May 2013. http://www.huffingtonpost.com. Accessed 8 July 2015.
Zimmerman, Michael E. "Heidegger and Deep Ecology." http://www.nchu.edu.tw/~hum/download/heidegger_deep_ecology.pdf. Accessed 7 June 2016.
———. "Nietzsche and Ecology: A Critical Inquiry." http://www.colorado.edu/ArtsSciences/CHA/profiles/zimmpdf/nietz_and_ecology.pdf. Accessed 6 June 2016.
———. "Rethinking the Heidegger-Deep Ecology Relationship." *Environmental Ethics* 15.3 (1993): 195–224.
Zimmerman, Virginia. *Excavating Victorians.* Albany: State U of New York P, 2009.
Žižek, Slavoj. *Living in the End Times.* London: Verso, 2010.

Index

B.L. Moore, *Ecological Literature and the Critique of Anthropocentrism*, DOI 10.1007/978-3-319-60738-2

D

The manufacturer's authorised representative in the EU is Springer Nature Customer Service Centre GmbH, Europaplatz 3, 69115 Heidelberg, Germany. If you have any concerns regarding our products, please contact ProductSafety@springernature.com

Printed and bound by CPI Group (UK) Ltd, Croydon, CR0 4YY
30/09/2026
02225649-0006